CAS Paper 13

Biofuel production strategies for UK agriculture

S P Carruthers and M R Jones

Centre for Agricultural Strategy
University of Reading
1 Earley Gate
Reading RG6 2AT

July 1983

ISBN 0 7049 0619 8
ISSN 0141 1330

Printed at the College of Estate Management, Reading

Originally established by the Nuffield Foundation in 1975, the Centre for Agricultural Strategy is a self-financing unit within the Faculty of Agriculture and Food at the University of Reading

STAFF

Director	Professor C R W Spedding
Personal Assistant to the Director	Miss A M Hoxey
Research Officers	Mr J L Jollans Mr A M M Thompson Mr R B Tranter
Research Associates	Mr S P Dimoline Miss D E E Ray
Secretary	Mrs M H Baker
Office Junior	Miss H E C Thomas
Honorary Research Fellows	Mr A Korbey Dr L J Peel Professor B G F Weitz
Consultant Econometrician (part-time)	Mr D Hallam (of the Department of Agricultural Economics and Management)
Computer Programmer (part-time)	Mrs V D McDougall

Preface

The Centre for Agricultural Strategy (CAS) has selected 'Energy in Agriculture' as one of its major projects, although the level of activity will be determined by the funding that the study attracts.

The reasons for doing so are fairly obvious, since agriculture is much concerned with the use of energy and, in developed countries, increasingly dependent upon 'support' energy (ie fossil fuels as distinct from solar radiation). However, it is the probable future cost of support energy that is the major cause for concern. If the cost of fossil fuels, and thus all those goods and services that use them, rises faster than other costs, then farmers will need to know what options are open to them, to reduce energy inputs, to increase the efficiency of energy use, and to produce their own.

One of the ways in which they might produce fuel is by growing biomass for conversion to usable liquids, solids or gases. This kind of 'biofuel' production has been studied, over the last few years, as part of a programme sponsored by the Department of Energy, Energy Technology Support Unit (ETSU) and we are most grateful to them for allowing us to draw freely on work they have financed, both at the University of Reading and at other institutions.

It is now timely to put together what has been learned in these studies, partly in relation to the work completed but also partly because of the current pre-occupation with overproduction of several major commodities.

Whatever means are finally adopted to discourage such overproduction, there have to be alternative uses for the resources (notably of land) released. Such alternatives have to produce what is wanted and not themselves result in overproduction and, in the long run, of course, they have to be profitable. Some opportunities already appear to be so, but many otherwise attractive biofuel production systems are not yet economic. Should (or when) they become so, they would represent alternative enterprises, probably for producing fuel to be used on site and thus capable of lowering costs. They would not contribute to over-production of food and would have the advantage of using existing farming skills: they would, however, represent an additional technology needing practical development.

Whether it should be encouraged or not has to be judged against what is so far known of biofuel production potentialities. This report is designed to present just that information.

C R W Spedding

Contents

ABBREVIATIONS

AD	Anaerobic Digestion
ADAS	Agricultural Development and Advisory Service
BABA	British Anaerobic and Biomass Association
BOD	Biochemical oxygen demand
CAS	Centre for Agricultural Strategy
CEC	Commission of the European Communities
COD	Chemical oxygen demand
CSO	Central Statistical Office
DART	Dartington Amenity Research Trust
DM	Dry matter
EA	Energy analysis
EC	European Community
ETSU	Energy Technology Support Unit
EW	England and Wales
GB	Great Britain
GER	Gross energy requirement
HMSO	Her Majesty's Stationery Office
ITE	Institute of Terrestrial Ecology
MAFF	Ministry of Agriculture, Fisheries and Food
MC	Moisture content
N	Nitrogen
na	Not available
N/A	Not applicable
NE	Net energy
NER	Net energy ratio
NFU	National Farmers Union
NIAB	National Institute of Agricultural Botany
SNG	Substitute natural gas
UK	United Kingdom
US	United States of America

UNITS

^oC	degree celsius	
kg	kilogramme	= 2.2046 lbs
ha	hectare	= 2.471 acres
h	hour	
J	joule	
m	metre	= 1.094 yards
t	tonne	= 0.984 tons
W	watt	
Wh	watt hour	
y	year	

MULTIPLES

k	kilo	10^3
M	mega	10^6
G	giga	10^9
T	tera	10^{12}
P	peta	10^{15}

ENERGY CONVERSION FACTORS

1PJ = 277.8 MWh (= 0.0379 million tonnes coal equivalent)

Foreword

This paper is a descriptive rather than prescriptive account of the potential for the production of biofuels in the UK. Technical potential is assessed, quantitatively, in terms of feedstock production, and comment made regarding the current 'state of the art' of the biomass technologies. The paper describes the economic and social factors likely to affect the development of biofuel production and discusses the implications for agriculture, employment and land use.

In view of the lack of reliable economic data and the effect of varying local conditions on economic viability, the paper does not attempt detailed quantitative assessments of economic potential either at a national or local level. Such studies would be an important next step in the exploration of biofuel potential. Similarly, the paper does not formulate a plan for a national biofuels policy, but provides a basis for doing so.

A number of studies have been carried out, in recent years, largely on feedstock production routes and conversion processes. This paper draws on many of these, aiming to synthesise the essential conclusions and, thus, present an overall picture. The paper draws attention to the potential of biofuels in the UK, but also emphasises the need for further research, development and demonstration if that potential is to be realised.

The authors would like to thank all those who provided them with many helpful comments and suggestions during the preparation of this paper

1 Introduction

The 'energy crisis' which followed the quadrupling of world oil prices in October 1973 led to a popular awareness of the limitations of fossil fuel supplies. This stimulated a surge of interest in energy conservation and in alternative energy sources.

The major alternatives to the conventional fossil fuels are nuclear power and renewables. The latter covers those sources of energy, such as wind, wave, direct solar, tidal and biomass, which do not draw on capital reserves, but are based, directly or indirectly, on the current income of solar energy or on gravitational energy; their rate of consumption will be comparable, therefore, to the rate of production.

Renewable energy resources are, in general, more evenly dispersed over the earth's surface than fossil fuels or uranium. Hence, they increase the scope for local, regional or national energy self-sufficiency. Additionally, renewable energies are not associated with environmental pollution, such as is caused by coal burning or liquefaction, or the kind of risks linked to nuclear power.

Fuels derived from biomass (ie biofuels) are not only potentially renewable, but are also sufficiently similar in origin to the fossil fuels to provide direct substitutes. In particular biomass has the potential to provide alternatives, such as ethanol, methanol or vegetable oils, to the petroleum-based liquid transport fuels. With the exception of the generation of hydrogen *via* photo-biological, photo-chemical or photoelectro-chemical processes (technologies at an early stage of development) no other renewable energy technology is capable of providing storable, chemical fuels. In the near future demand is likely to be, primarily, for such high density chemical fuels rather than the low grade heat, mechanical power or electricity provided by the other renewable energy technologies.

Because of this capacity to provide storable chemical fuels, biomass can also be used in a non-renewable manner. Much of the developing world is suffering, at present, a severe 'energy crisis' as firewood has been used at a rate far in excess of its regeneration, and natural forests have been depleted.

In the wake of the energy crisis, biofuels, in common with a number of other renewables, were seen as an attractive option and became the subject of much research and development. Most of this has centred on the technologies for converting biomass into fuels (ie conversion technologies); research, development and, more recently, demonstration of several different technologies has aimed at improving output, efficiency and reliability and at lowering costs. Some research has also been carried out to assess the quantity and quality of resources available in the UK for biofuel production and on the cultivation of energy crops, notably trees.

Less attention has been paid to the implementation of these technologies and production systems; to the overall value of biofuels in relation to national energy supplies and other national objectives; and to the implications for, and impact on, those sectors of industry in which biofuels would be produced or which would compete for the same resources.

Implementation of biofuel production systems will depend on:

(i) the technical feasibility of the conversion process in terms of efficiency, reliability and the availability of the hardware;

(ii) the quantity, quality and ease of availability of feedstocks, or land to produce energy crops, for these processes;

(iii) economic factors, including the costs of feedstock production, conversion, storage, transport and utilisation, the current prices of conventional fuels and the degree of risk or uncertainty associated with the system;

(iv) social factors, including the ease of integration with other related activities and reactions to the technologies or other social changes related to the implementation of biofuel systems;

(v) government policy.

The overall potential extent of implementation will determine the importance of biofuels on a national scale, and the means of implementation will determine the extent and nature of the impact on current practice. An understanding of these effects would facilitate the formulation of government policy in relation to biofuels.

The essential resource for biofuel production is biomass which, other than in the case of marine or aquatic systems, needs land for its production. The 'aquatic' resource in the UK is fairly limited; marine biofuel systems have considerable potential, but a detailed discussion of them is beyond the scope of this paper. Agriculture, defined here as the management of land to produce biological commodities (hence including forestry) will, therefore, be the context for most biofuel production. Any major biofuels programme will thus have strong implications for agriculture.

This paper aims to:

(i) review the biomass conversion technologies and fuel crop production systems applicable to the UK and comment on the current 'state of the art';

(ii) assess the overall potential for biofuel production from UK agriculture and to consider possible changes to increase this potential;

(iii) discuss the factors affecting the implementation of biofuel production systems;

(iv) consider the implications of the implementation of biofuel production systems for agriculture and land use, employment, energy, food and timber supplies and government policy;

(v) make recommendations for further research.

In Chapter 2 of this paper the biomass conversion technologies are reviewed: the types of feedstocks required are defined, the processes and products briefly described and some assessment made of the current 'state of the art'. The UK biomass system is then described and various possible strategies for producing biomass for fuel production are presented, along with estimates of their potential output, in Chapter 3. Chapter 2, therefore, describes the processes available and the type of feedstock required, while Chapter 3 describes means by which specific feedstocks might be made available for these processes. A number of these involve the release of land for growing fuel crops; Chapter 4, therefore, discusses the choice of crop for such situations. In Chapter 5 the economic, technical, social and organisational, and legal and institutional factors affecting the implementation of the biofuel production systems described in the previous chapters are discussed. The implications of biofuel production for land use, food and timber supplies, the agricultural industry and employment, and the role of government policy in relation to biofuel production are considered. The main conclusions of the paper are discussed in Chapter 6 and recommendations made for further research. Appendices describe the model used to formulate biofuel production strategies, discuss energy analyses and exemplify the economics of biofuel production *via* several case studies. Other appendices include a directory of UK energy from biomass organisations and a glossary of terms.

2 Conversion technologies

Fuels may be produced from biomass using a number of conversion processes. These are illustrated in Table 2.1. Most conversion processes are fairly specific in the type of feedstock they require. The type of feedstock available is thus likely to determine the type of fuel which can be produced. The different conversion processes are discussed in this chapter and the potential feedstocks are considered in Chapters 3 and 4.

DIRECT COMBUSTION

Direct combustion of biomass to produce heat for cooking, hot water or steam generation is a well established technology. Traditional combustion of biofuels in stoves has low efficiency (30%) compared to modern boilers which may achieve 65% efficiency. Efficiencies as high as 80% may be possible in large (9MW) straw-burning plants with automatic stoking (Tougaard Pedersen, 1982) and, using a gas scrubber as the heat exchanger, Kofoed Neilsen & Neilsen (1983) report efficiencies of 86–96%, as well as a halving of particle emissions.

Feedstocks for direct combustion should have moisture contents of no more than 15%, though wetter feedstocks may be burned with consequent loss of thermal efficiency to evaporate excess water. Direct combustion therefore requires drying and weatherproof storage of feedstocks, which will increase costs. Other treatments such as compaction of straw and woodwastes into briquettes or pellets to improve energy density, and chopping of straw or chipping of wood to improve handling characteristics may also be necessary. The labour requirements, especially for manually stoked boilers, may be quite high — as much as 6 man-hours per tonne (Bishop, 1980), and problems may occur from smoke emissions if combustion efficiency is poor. Direct combustion is, however, a widely used technique: it can use a variety of feedstocks which are readily available on many farms and can be operated efficiently at a small scale.

Table 2.1
Biomass conversion technologies

Process	Feedstock quality	Typical feedstocks	Fuel produced	Liquid fuels?[1]
Direct combustion	dry	straw wood	(heat)	N
Pyrolysis	dry	wood	oil char gases	Y
Gasification	dry	wood dry residues	gas	(Y)
Direct liquefaction		wood	oils	Y
Aerobic decomposition	wet	animal waste	(low grade heat)	N
Anaerobic digestion	wet	animal waste green crops	biogas	(Y)
Fermentation	high sugar content	sugar beet	ethanol	Y
Fermentation / hydrolysis }	starchy	potatoes, grain	ethanol	Y
Cellulosic fermentation }	cellulosic	green crops	ethanol	Y
Oil extraction	high oil content	oilseed rape	vegetable oils	Y

Note:
1 N = No; Y = Yes; (Y) = indirectly to methanol.

Present use of direct combustion on farms is probably the most developed and economically attractive application of renewable energy in UK agriculture. Wood and straw burning furnaces are commercially available, and have been installed on several hundred farms. The primary application of the heat output has been for domestic heating, though some systems are used to heat pig units, glasshouses and other enterprises. The present trend in commercial straw burner technology is toward minimising the processing required for efficient combustion

(Teisen, 1982), for example by using batch burning of large bales with variable air supply, and heat storage.

On a national scale direct combustion of agricultural residues may have potential as an auxiliary fuel for electricity generation (Hitzhusen & Abdallah, 1980; English *et al*, 1981; Jenkins *et al*, 1981). UK opportunities for such uses, however, may be limited by the availability of sufficient intensity of residue production to supply an economically-scaled plant, though Thring & Crookes (1981) suggest that local power stations of a few hundred kW could be straw-fired if fuel prices rise and field burning is restricted. Intensive silviculture could provide power station fuels, though Rose (1977) estimates that a 150MW generating plant would require a catchment area of 66.34 kha with a 15 year rotation. Natural biomass in the form of peat is at present used as power station fuel in Eire.

THERMAL PROCESSES

A number of thermal processes are available for the conversion of biomass to fuels. These are illustrated in Figure 2.1.

Figure 2.1

Possible thermal routes for the conversion of biomass to fuels

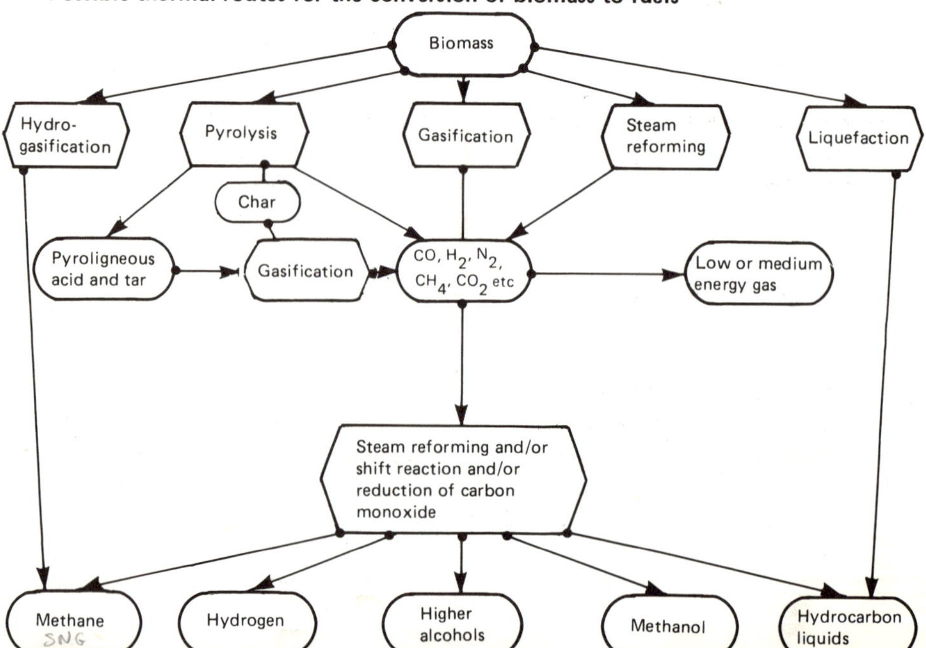

Pyrolysis

Pyrolysis is the destructive distillation of biomass in the absence of air. The feedstock should generally be dry (less than 15% MC) as for direct combustion, though biomass up to 40% MC may be used. Pyrolysis occurs at 200–1100 °C to produce char, liquid distillate and mixed gases. The liquid distillate is a complex mixture consisting of two phases – a phenolic tar and an acidic aqueous phase which may cause corrosion problems, in the reaction vessel or when used as a fuel, unless neutralised. The gases formed are also a mixture with a heating value of 10.5–15.0 MJ per m^3. Conditions may be optimised to produce 'synthesis gas' ($CO + H_2$) as a preliminary stage to methanol production. The relative proportions of char, distillate and gases are quite variable depending on the pyrolysis temperature and rate of heating, which will be affected by the feedstock type and moisture content. A number of pyrolysis techniques have been designed, giving different yields of the three products. A typical yield from 'slow pyrolysis' may be 35% char, 35% distillate and 22% gases on an energy basis (Bliss & Blake, 1977), whereas with high heat transfer rates 'fast pyrolysis' may lead to practically no char at all (Reed, 1981).

The char has a heating value of 30.0 MJ per kg (*cf* coal 28.8 MJ per kg) and is therefore a useful fuel, though it may also have potential in steel-making, as activated carbon, or for gas purification. The liquid may be used directly, or further distilled, to produce a fuel or chemical feedstock. The gases also have potential as a fuel, and can be burned in a gas generator to produce electricity. Pyrolysis thus produces a variety of energy concentrated fuels, some of which are storable. At present, product yields are not entirely predictable and this may be a constraint on the application of the process.

Pyrolysis is a relatively simple technology, which could be adopted at the farm scale, though as yet no plants are in operation. Alternatively, Linneborn & Tremus (1981) describe a mobile pyrolysis plant which could be used by a group of farms. Large-scale pyrolysis has been developed for treatment of municipal wastes and a number of commercial processes are available (Bungay, 1981), though only one plant is in operation in the UK, operating on rubber tyres.

Gasification

Gasification involves the thermal conversion of biomass to gases. There are three alternative gasification processes: air gasification, producing a low energy gas containing 42% nitrogen; oxygen gasification, producing 'synthesis gas'; and gasification with hydrogen to produce 'substitute natural gas' (SNG). The addition of steam or the presence of water in the feedstock improves yields of synthesis gas and reduces byproducts. Gasification can therefore utilise wetter feedstocks (20% MC) than other thermal processes, and with the 'MINO' process, which has catalytic reforming of tars (Blomkvist *et al*, 1983) biomass with a moisture content of 50% may be used.

Synthesis gas is a versatile product. It may be burned directly (energy value is 10–20 MJ per m^3) to provide heat, for example for grain drying, used in gas engines to provide mechanical energy or to generate electricity, or may be used to produce methanol which can be used as a liquid transport fuel. There are problems to the use of methanol as a transport fuel, however, largely due to its toxicity and problems of corrosion.

A variety of gasifier designs are being studied, the main variations being co- and counter-current and fixed and fluidised bed designs. Fixed bed counter-current gasifiers are the traditional design, but are limited in their range of feed-stock, generally requiring dry chipped wood. Loose feedstocks such as straw or sawdust require updraught gasification, which may have problems of tar formation. Fluidised bed systems have been more recently developed and may accept a wider range of feedstocks.

Farm-scale applications for methanol production are likely to be limited due to the complexity of the process, which requires high gas pressures, and oxygen supplies. Large-scale methanol production from wood, however, may be an attractive route to liquid fuels in temperate regions. Chrysostome & Lemasle (1983) report results of a pilot, pressurised oxygen-blown gasifier with a 60 t per day throughput, with an efficiency of 1 t of methanol from 2 t of dry wood. For the UK, Brandon *et al* (1983) estimate that a 1 ktDM per day gasification plant (about the size of some pulp mills) could be supplied from wood energy plantations within a 20 mile radius.

The production of low energy gases from air gasifiers may have potential for smaller scale (0.1–20.0 MW) combined heat and power generators (Van Swaaij, 1981), for which Zerbin (1983) quotes an efficiency of 1.24 kg air-dry wood per kWh electricity. Farm-scale gasifiers have been developed in the US (Loewer *et al*, 1982) operating on maize cobs to dry grain. In the UK there are no commercial gasification plants in operation, though two pilot plants have been established (Van Swaaij & Beenackers, 1982; Wilson, 1982).

Direct liquefaction

Biomass may be directly converted to liquid fuels by catalytic liquefaction. Burton *et al* (1983) describe a process for the catalytic hydrogenation of wood to produce a mixture of liquids suitable as a fuel and/or chemical feedstock. Direct liquefaction may also be applied to wet cellulosic and lignocellulosic wastes by carboxylolysis (Bliss & Blake, 1977). CO and steam are used to hydrogenate biomass in the presence of sodium carbonate catalyst to produce oils or methane-rich gases. The process has a maximum efficiency of 50%. It is well suited as a treatment for wood wastes as lignin as well as cellulose is converted.

BIOLOGICAL PROCESSES

Aerobic decomposition

Aerobic decomposition of animal wastes is exothermic, and has thus attracted some interest as a potential source of low grade heat. Aerobic treatment of animal wastes is already carried out on some farms as a pollution control measure, though the process is rarely optimised for heat production. Tjernshaugen (1981) describes a system using mechanical aeration of animal slurry to generate heat which could be transferred via a heat exchanger for domestic heating. The odour of aerobically treated waste is less than untreated material, though odour output may increase during treatment. After aeration manure handling is easier and pathogens are reduced. Traditional handling of solid animal manures by composting may also be adapted for heat production with an efficiency of 17–34% (Schuchardt, 1983). Heat extraction from aerobic decomposition is thus a simple and robust method of producing low grade heat and may be readily integrated with existing farming practice.

Anaerobic digestion

Anaerobic digestion can be applied to any wet feedstock (85–95% MC), though its main development has been as a treatment for animal and human wastes. The technique is effective in lowering the biochemical oxygen demand (BOD) and chemical oxygen demand (COD), reduces odour problems and pathogens and leaves a manageable slurry which retains much of the plant nutrient value of the original waste. Anaerobic digestion yields a mixture of carbon dioxide and methane (biogas) typically 30% CO_2, 70% CH_4 with an energy value of about 25 MJ per m^3.

The efficiency of gas production depends on the type of feedstock: pig slurry may yield 65% of gross thermal value as gas, fresh grass 60%, and cattle slurry 33%. In UK conditions, especially in winter, a proportion of the gas produced must be used to maintain the digester contents near to the optimum 35 °C to sustain gas output. Further energy may be needed to stir the contents to avoid crust formation, thus net efficiencies are likely to be lower than these figures. Not all anaerobic digestion systems operate at mesophilic temperatures. Cryophilic systems operating at 15 °C with long (100 day) retention times are being studied (Kaufmann *et al*, 1982; Kopiske & Eggersgluß, 1983), as well as those operating in the thermophilic range (45–70 °C) (Lettinga, 1981).

Apart from the conventional fully-stirred contact digester, there are several new digester designs at the experimental stage. Plug-flow digesters (Stafford & Etheridge, 1982) which do not necessarily require mechanical mixing have advantages over conventional stirred digesters in that substrate concentration is maximised and all elements of the feed have the same retention time. The design of Hawkes *et al* (1981) minimises the problems of scum formation, whilst retaining the simplicity and reduced energy requirements of the plug-flow design. US experience (Scheller, 1983) indicates a two- to sevenyear payback time for a plug-flow digester converting the gas to electricity. Fixed and fluidised bed anaerobic filters (Donellan & Newall, 1982; Heijnen, 1983) in which the methanogenic micro-organisms are suspended in a flow of substrate, help to reduce the retention times. Two-stage digesters (Rijkens, 1981; Barry *et al,* 1982) may be useful in treating cellulosic and high-solid wastes. The separation of liquefaction and methanogenesis allows higher loading rates and lower retention times, without increasing system complexity, and can operate on a wide variety of feedstocks. Hydraulic digesters (Stafford & Etheridge, 1982) have potential for high rate, low retention times even with wastes containing relatively high suspended solids.

Anaerobic digesters have been used at several sewage works in the UK for some years, and and a number of different digester designs are commercially available for animal waste treatment. Twelve biogas plants have already been installed on farms in the UK and nine pilot-scale plants are in operation (Demuynck *et al,* 1983). The biogas produced may be burnt to provide space heat or hot water, converted to electricity *via* a modified diesel generator or, if scrubbed and compressed, used as a vehicle fuel (Held & König, 1983; Picken, 1983), or partially oxidised to methanol.

Fermentation

Ethanol produced by fermentation of plant sugars has been used as a beverage for over 10 000 years. It is also a valuable transport fuel with an energy content of 28 MJ per kg and an energy density of 28 GJ per m^3 *(cf* 46 MJ per kg and 35 GJ per m^3 for petrol). The production of high grade (95% pure) alcohol, suitable for fuel use, using conventional techniques requires energy intensive distillation of the dilute alcohol solution (beer) left after yeast fermentation and disposal of the distillery effluent, which is 10–14 times the ethanol volume. Raw materials make up the majority of total production costs at present, thus improvements in process equipment costs are unlikely to significantly alter production economics, though energy efficiencies and conversion costs could be improved. At present energetic efficiencies are low even using sugary feedstocks as Table 2.2 shows. It is apparent that the use of crop byproducts, such as maize stover, for process heat can help improve net energy ratios. In the UK there are few crops with sufficiently high sugar contents to be suitable feedstocks for fuel ethanol production. Higher yielding starchy, cellulosic and lignocellulosic crops could help to reduce feedstock costs. These may be enzymatically or chemically hydrolysed to provide fermentable substrates (Beldman *et al,* 1982; Franz *et al,*

1983). An alternative route for production of ethanol from such feedstocks is by direct cellulosic fermentation (Smith, 1980), though this process is still at a very early stage of development.

Ethanol is a versatile fuel which can be used pure, or mixed with petrol (US 'Gasohol'), as a transport fuel. Addition of 10–15% alcohol to petrol improves the combustion efficiency, raises octane ratings such that lead additives are unnecessary (Emery & Kent, 1979), and requires little modification of conventional engines. Engines modified to run on pure alcohol have been successfully developed (Bernhardt, 1981) and are widely used in Brazil.

On-farm production of illegal 'moonshine' alcohol has a long history, but more recently there has been renewed interest in systems specifically designed for farm-scale production of fuel alcohol. These are mainly at the pilot plant stage. Dodd & Roberts (1981) and Pile *et al* (1981) describe two such systems which are designed to be simple, and energy efficient. Studies by Buckland & Buick (1980), and Wonder & Simpson (1982) for Australia, and Lockeretz (1982) for the US, however, indicate that farm-scale production is uneconomic at present. Even though the energetics and economics of fuel ethanol production are generally not very satisfactory, on particular farms with, for example, cheap feedstock supplies, the advantage of 'home production' of liquid transport fuels may make it worthwhile.

Alternative industrial processes for ethanol production, offering improved energy efficiencies, are also being developed, but this has been mainly at the cost of increased system complexity (Essien & Pyle, 1982). Research is being conducted at all three stages of ethanol production (ie fermentation, product recovery and waste treatment) to improve energy efficiencies. Fermentation research has concentrated on four areas: increasing ethanol productivity to reduce fermenter volume; improving sugar conversion efficiency; increasing ethanol concentration in the 'beer' to reduce distillation energy and effluent treatment; and development of simpler and cheaper processes. Conventional fermentation is a batch process; continuous fermentation is relatively simple and offers higher ethanol productivity, though product inhibition and cell washout may be a problem, without addition of cell recycle. Immobilised cell systems allow high cell densities and very high productivities though the absence of an effective degasification system reduces its industrial potential. Extractive fermentation techniques are also being studied as they reduce ethanol inhibition, but they are more complex and less reliable than conventional systems.

At the distillation stage, vapour recompression may be used to eliminate process steam requirements, and might be attractive if electricity was cheaply available. Freeze crystallisation reduces energy requirements but still requires further distillation to produce 95% alcohol. Liquid-liquid extraction and adsorption on to cornstarch are other novel processes under consideration.

Effluent disposal requirements may be reduced by use of more concentrated feeds at the distillation stage. The only disposal process for which there is a potential net energy gain would be anaerobic digestion of the effluent. The investment required for an anaerobic digester associated with an ethanol production plant may be nearly as large as for the ethanol plant itself (Costa Ribeiro & Castello Branco, 1981).

Table 2.2

Energy balance of ethanol production from temperate crops

Crop	Net energy ratio 1	Net energy ratio 2	Source
Maize	na	2.33	Lockeretz (1982)
Irrigated maize	1.08	2.41	Ofoli & Stout (1981)
Dryland maize	1.16	2.81	Ofoli & Stout (1981)
Sugar beet	na	1.76	Parisi (1983)
Sugar beet	1.67	2.44	Revuz & Daujat (1983)
Maize	1.25	na	Scheller (1983b)

1 non-renewable fuels used for process energy

2 biofuels used for process energy

MECHANICAL PROCESSES

Oil extraction

Vegetable oils which can be used as a fuel for diesel engines may be produced by mechanical or solvent extraction. Cheap screw presses are available for small-scale oil extraction, but leave about 14% oil in the residue. Larger mechanical presses are more efficient (5% oil in the residue), but industrial-scale oil production is likely to use solvent extraction in conjunction with pre-pressing to achieve 1% oil in the residue (Radley & Garrod, 1981). Feedstock costs are a significant proportion of the total (77–80%), thus improvements in process efficiency are unlikely to overcome the twofold higher price of these oils in comparison with diesel (Stewart *et al,* 1981). For on-farm oil production a major item of cost may also be the capital investment for the press; by comparison, labour and fuel inputs are only a small item (Wonder & Simpson, 1982). Pure vegetable oils cannot be used satisfactorily in unmodified diesel engines, thus some sort of pre-processing is likely to be necessary if they are to be used in conventional engines. The protein-rich residue left after oil extraction can be used for animal feed, and thus may offset some of the production costs. Effluent disposal, however, may be an additional cost.

3 Feedstock resources

Whatever advances are made in relation to biomass conversion processes, they will be of little value if feedstocks are not available in sufficient quantities and at appropriate costs. Dr P Chartier (Directeur Scientifique de l'Agence Française pour la Maitrise de l'Energie and Project Leader Energy from Biomass of the Commission of the European Community), summing up the 2nd EC Energy from Biomass conference considered the availability of resources (ie feedstock quantity, costs and seasonality) to be the major limiting factor in all the scenarios (except waste use) being examined by the EC Biomass Research Programme.

A number of studies have been carried out on biofuel production resources in the UK and these are summarised in this chapter. These studies fall into three broad categories:

(i) assessments of the quality, quantity and availability of wastes and by-products (Ader & Buck, 1979; Plaskett, 1980; Larkin *et al,* 1981; Mitchell & Matthews, 1981) or of unharvested natural vegetation (Newman, 1979; Lawson *et al,* 1980, 1983);

(ii) studies of the productivity of various types of energy crop including trees (McLain, 1983b; Mitchell & Matthews, 1983; Pearce, 1983; Stott *et al,* 1983), agricultural crops (Spedding & Carruthers, 1980; Carruthers, 1981a, 1982; Spedding *et al,* 1979b, 1980) and indigenous or introduced species (Callaghan *et al,* 1981);

(iii) estimates of the areas of land available for some of these energy crops (Dennington & Chadwick, 1979; Spedding *et al,* 1979b; Mitchell *et al,* 1983).

In the UK biomass is produced largely within agriculture (which here includes forestry and horticulture) towards the provision of consumable products or, to a small extent, amenity. Table 3.1 draws attention to the high proportion of UK land (87%) that falls into this category and indicates that there is little spare land available for the unlimited production of biofuel feedstocks.

Table 3.1
An estimate of land uses in the UK, 1976

Land use	kha	%
Crops and fallow	4 802	19.9
Temporary grass	2 314	9.6
Permanent grass	4 950	20.5
Rough grazing	6 768	28.1
Other land[1]	312	1.3
Forestry and woodland	1 783	7.4
Total agriculture	20 929	86.8
Urban	1 845	7.7
Miscellaneous	1 326	5.5
Total land	24 100	100.0

1 Includes 'woodland ancillary to farming', farm roads and buildings.

Source: Adapted from Tranter (1983).

Although the limited land resource and the demand for food and timber constrain the production of biofuels, they do not, necessarily, prevent it; a number of ways of making resources available for biofuel production within the UK agricultural system are possible. Wastes, byproducts and unharvested natural vegetation could provide suitable feedstocks for a number of biomass conversion processes, some areas of unexploited land may be suitable for fuel cropping and further areas of marginal land may be more profitable under energy crops than the current enterprise. The opportunities for fuel cropping might be increased by improvement in the efficiency of crop and animal production, and circumstances may arise where an increased national commitment to biofuel production is considered particularly desirable.

As a basis for the examination of these possibilities this paper uses a conceptual model of the UK biomass system. This model, a simple hydrodynamic analogue of solar energy flow, is illustrated in Figure 3.1, and explained in more detail in Appendix 1. From a study of the model five basic strategies for increasing the amount of biomass available for fuel production emerge:

(i) decreasing the amount of 'unused' biomass energy (valves 8 and 11);
(ii) increasing total primary production (valve 1);
(iii) decreasing the input to secondary production (valve 3);
(iv) decreasing outputs (valves 4, 5 and 6);
(v) increasing imports (valve 2).

Figure 3.1
Energy flow through the UK biomass system (see Appendix I)

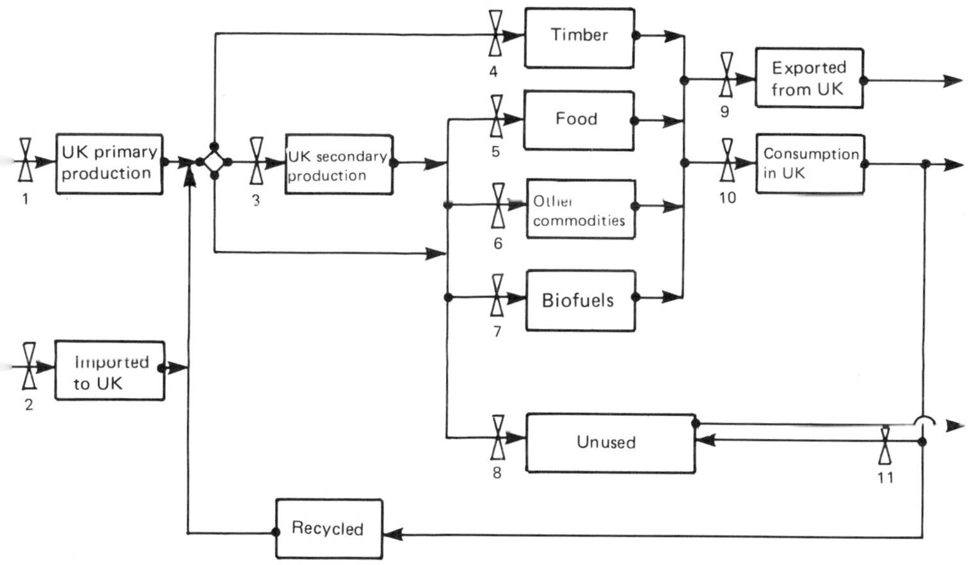

Each of these strategies, and the various resources which would be made available by applying them, are now considered. Throughout the rest of this chapter estimates of the potential of a number of possible opportunities are presented; these estimates are expressed both in terms of energy output in PJ per year and as percentages of national primary energy consumption in 1981. The latter serves only as a guideline as changes in national energy consumption will alter the apparent potential of any opportunity when expressed in this way.

National primary energy consumption is equivalent to the total inland consumption of primary fuels. According to the Department of Energy (1982) this refers to the 'total input into the economy of primary fuels and their equivalents' (coal, petroleum, natural gas, nuclear and hydro-electricity) and the measure therefore includes: energy used and lost in the conversion of primary fuels to secondary fuels; energy lost in the distribution of fuels; and energy conversion losses by final users. Full definitions of these fuels and other measures of energy consumption appear in Department of Energy (1982).

The total inland consumption of primary fuels in the UK has fluctuated around an average of 8869 PJ per year in the last 10 years (Table 3.2) and decreased by 11% from 1979 to 1981 (the lowest value since 1968). Greater emphasis on energy saving and more widespread adoption of energy conservation measures (some of which may be in the development phase at present) may significantly reduce total energy consumption in the future.

Table 3.2

Total UK inland consumption of primary fuels since 1972

Year	PJ[1]
1972	8 915
1973	9 324
1974	8 902
1975	8 567
1976	8 699
1977	8 926
1978	8 937
1979	9 382
1980	8 670
1981	8 369

1 Original figures in million tonnes coal or coal equivalent. Assuming conversion to PJ at 26.377 PJ per million tonnes coal or coal equivalent.

Source: Adapted from Department of Energy (1982)

It should also be noted that although the output estimates are all expressed in the same energy units (ie PJ) and are all expressed as a proportion of total UK primary energy consumption (itself comprised of a number of different types of fuel), they represent a number of types of fuel which will differ in their quality and utility. Additionally, the estimates presented do not indicate the magnitude of, or likely differences in, the fuel inputs to production and conversion of these fuels, either in comparison with one another or with conventional fuels. Both fuel quality and support energy inputs will be reflected in an economic analysis and the latter is also indicated in an energetic analysis. These forms of analysis are discussed in a subsequent chapter.

DECREASING THE AMOUNT OF 'UNUSED' BIOMASS ENERGY

Unused biomass refers to the wastes and byproducts of pre-consumption processes (ie crop and animal production, food packaging and processing etc) and post-consumption wastes, and to unharvested natural or semi-natural vegetation. A decrease in the quantity of unused material may also be achieved by system changes, designed to exploit produced biomass more fully. A classification of potential resources for biofuel production made available by this strategy appears in Table 3.3, together with a summary of estimates of potential energy output.

Urban wastes

The term 'urban wastes' here refers to those organic solid wastes derived from general consumer use or from manufacturing processes. The former category includes industrial and commercial refuse, domestic and trade refuse and specific items such as scrap tyres, spent lubricating oil and waste paper; the latter refers to residues from manufacturing processes which cannot be recycled back to the same or a related process. A full account of the potential of these wastes as a fuel source appears in Ader & Buck (1979); figures for urban waste output in Table 3.3 have been drawn from this document. It should be noted that the quoted figures for all wastes, except sewage sludge, are means of estimated ranges of the energy content on a dry and ash-free basis of the total arisings requiring disposal in 1976—77. (Sewage sludge output is based on anaerobic digestion to biogas at an efficiency of 8.5 GJ methane per dry tonne of sewage.) The heterogeneous nature and the fairly high proportion of inorganic materials and water in refuse would make the energy yield on combustion lower than the quoted figures. Additionally, only a proportion of the total production could be feasibly and profitably utilised. Ader & Buck (1979) suggest values for this proportion and also project their estimates to the year 2000.

Table 3.3

Estimates of potential annual UK energy output of various means of decreasing the amount of 'unused' biomass energy

Resource	Energy output (PJ)	
Industrial and commercial refuse	220.00[1]	
Domestic and trade refuse	210.00[1]	
Oil and tyres	25.00[1]	
Sewage sludge	20.00[1]	
Process wastes	6.00[1]	
Total urban wastes		481.00
Dry crop residues	147.90[2]	
Wet crop residues	14.87[2]	
Animal wastes	50.92[2]	
Horticultural wastes	0.91[2]	
Wood residues	21.64[3]	
Early thinnings	6.00[4]	
Total agricultural wastes		242.24
Total wastes and byproducts		723.24
Rural natural vegetation	58.68[5]	
Farm woodland	15.27[6]	
Urban vegetation	16.50[5]	
Aquatic and marine vegetation	4.29[7]	
Total unharvested vegetation		94.74
Total		817.98
% 1981 UK primary energy consumption[8]		9.77
System changes – dual purpose oilseed rape		93.20
% 1981 UK primary energy consumption[8]		1.11

1 Estimated total production requiring disposal 1976/1977
2 Estimated available production 1976
3 1977
4 1980
5 1975–1979
6 1979
7 Source year not specified
8 8 369 PJ

Sources: Adapted from Ader & Buck (1979); Newman (1979); Spedding *et al* (1979a); Spedding *et al* (1979b); MAFF (1980); Larkin *et al* (1981); Mitchell & Matthews (1981); Lawson & Callaghan (in press).

Industrial and commercial refuse
Industrial and commercial refuse includes packing materials, general rubbish, food wastes and other products normally arising within factory or commercial premises and handled by waste contractors. The composition is very variable and probably only about 30–40% is combustible. However, the development of refuse-derived fuel may provide an incentive for the segregation of combustible materials at source.

Domestic and trade refuse
Domestic and trade refuse refers to those materials collected by Local Authorities and disposed of by County Waste Disposal Authorities. As with industrial refuse composition is variable, but the organic combustible content may be higher (*c*60%), making domestic refuse a potentially valuable fuel source. Segregation of combustibles at source would, however, be difficult to effect, and exploitation of both types of refuse assumes successful development of combustion and energy utilisation systems.

Oil and tyres
Spent lubricating oil and scrap tyres are small but potentially easily available resources for fuel production.

Sewage sludge
Sewage sludge is already used as a feedstock for methane production via anaerobic digestion at some larger sewage works. There is scope for greater exploitation of this resource, but attention needs to be paid to the profitable utilisation of the gas.

Process wastes
The food and drink industry gives rise to substantial quantities of residues, but practically all are saleable for further conversion to other foods or for use in animal feed (Ader & Buck, 1979), leaving only a small quantity available for fuel production. However, Fallows & Wheelock (1982) suggest that food processing wastes, although potentially available as animal feed etc, are not actually being utilised efficiently, but rather causing environmental pollution or considerable expense to prevent it. Use as a biofuel feedstock may provide a realistic and, in some situations, a more feasible outlet. Other industrial sectors which may give rise to process wastes which may be exploitable as fuel feedstocks include the manufacture of organic chemicals, paper and board and the textile industry.

Dry crop residues

A number of crops grown in the UK provide dry residues for direct combustion or biological or thermal processing to liquid or gaseous fuels. Cereal straw, for example, could supply all agriculture's heating demand if burnt as a fuel (White, 1980). An estimate of the annual production of dry crop residues in the UK is presented in Table 3.4.

Table 3.4

Annual production and availability of dry crop residues, UK, 1976

Crop	Production		
	Total	Available[1]	
	kt DM	kt DM	PJ[2]
Wheat	4 371	4 109	71.09
Barley	5 583	3 661	64.07
Oats	689	379	6.63
Mixed corn	71	71	1.24
Rye	25	25	0.45
Field beans	132	132	2.30
Dried peas	87	9	0.16
Oilseed rape	109	109	1.92
Total	11 067	8 495	147.90
% 1981 UK primary energy consumption[3]			1.77

1 Available production refers to material which is not used or likely to be used as animal feed.

2 Energy contents adjusted to allow for latent heat required to remove water when straw is burnt.

3 8 369 PJ.

Source: Adapted from Larkin *et al* (1981).

The yield of straw in cereals is determined by the species, variety, environmental factors and cutting regime. A change to longer-strawed varieties would increase straw yield, although this would also increase the risk of lodging. Additionally, a low grain-to-straw ratio is not necessarily associated with low grain yield and there is scope for improving cereal straw yields by probably 50% through breeding and selection. Straw yields will also be affected by environmental factors, especially nitrogen level; the output of straw also shows noticeable variation from season to season (Smith *et al*, 1975). In general, crops which give good grain yields also give heavy straw yields.

The total above-ground biological yield in cereals is appreciably higher than the sum of average grain and straw yields. The dry matter yield of above-ground plant parts other than grain in winter wheat falls in the range of 8–11 t per ha (Smith *et al*, 1975). The cutting regime is, therefore, a very significant determinant of straw yields, and straw yields of combine-harvested cereals are probably well below the recoverable potential.

The combine cutting height is determined by the unevenness of the ground, the amount of green material in the crop and straw handling policy (ie baling, ploughing or burning); a higher cutting height usually promotes a faster throughput. Lowering the cutter bar or stubble harvesting with a forage harvester would increase straw yields; it is also claimed that the technique of whole crop harvesting recovers 100% more straw, as well as the light chaff fraction, than conventional handling (Lucas, 1978). In this method the crop is harvested at an earlier stage and dried at a central processing plant, and the extra straw recovered needs to be offset against the extra energy needed for drying.

Wet crop residues

Wet, green crop residues provide a feedstock suitable for anaerobic digestion to produce methane. An estimate of wet crop residue production appears in Table 3.5; sugar beet tops make by far the largest contribution. Improved management aimed at increasing root yields in sugar beet and potatoes could lead to equivalent increases in byproduct output; sugar beet yields in NIAB trials are 30% greater than UK average yields, potato yields 75% greater.

Wet crop residues could be harvested by forage harvester and, if necessary, stored as silage. This would increase costs and involve dry matter losses, but would enhance the efficiency of anaerobic digestion (Stewart, 1980).

Horticultural wastes

National production of horticultural wastes is small, but they may be of local importance. The production of some horticultural wastes is estimated in Table 3.6. Tomato, carnation and hop residues have a low dry matter content and could provide a feedstock for anaerobic digestion; in the case of the first two the glasshouse system may provide an outlet for the methane and digester residue. Anaerobic digestion may also aid in the control of plant pathogens (Stafford, 1982). Orchards and raspberries yield a drier residue.

Table 3.5
Annual production and availability of wet crop residues, UK, 1976

Crop	Production		
	Total	Available[1]	
	kt DM	kt DM	PJ[2]
Sugar beet	1 984	810	7.90
Potatoes	369	362	3.53
Peas for processing	191	179	1.75
Beans for stockfeeding	132	132	1.29
Brassicas	128	121	1.18
Other legumes	66	63	0.61
Carrots and parsnips	62	56	0.55
Turnips and swedes	6	6	0.06
Total	2 040	1 729	14.87
% 1981 UK primary energy consumption[3]			0.17

1 Available production refers to material not used or likely to be used as animal feed.

2 Assuming net methane yield *via* anaerobic digestion of 9.75 GJ per tonne DM.

3 8 369 PJ.

Source: Adapted from Larkin *et al* (1981).

Animal wastes

Animal manures provide a suitable feedstock for anaerobic digestion to produce methane. The total output of wastes is determined by the number of animals in the country and the mean output per animal, but only manure from housed animals is potentially recoverable; a proportion of cattle waste and nearly all pig and poultry manures are, therefore, recoverable and in fact, require disposal. As disposal costs increase and livestock units become more intensive, the conversion of waste into useful energy would seem attractive. An estimate of animal waste output appears in Table 3.7.

Wood residues

Residues from forestry and wood processing provide potential resources for direct combustion or thermal processing. A proportion of the latter are currently used for pulping, chipboard, animal bedding and other uses; forestry residues, comprising 45% of the total biomass, are normally left in the forest. An appreciable proportion of these residues is potentially recoverable. Table 3.8 presents estimates for the production of forestry residues in 1977 and projections for 2000. Owing to the age structure of British forests and the intensive planting programme since 1950 the quantity of forest residues arising will more than double by 2000 (Mitchell & Matthews, 1981). Additionally, improvement in residue recovery will render a greater proportion of these residues available. Both these effects are reflected in Table 3.8. Further expansion of UK forestry will generate further residues in due course.

Early forest thinnings

'Thinning is an essential part of the management of British plantations because the prime object of silviculture is to produce saw logs. At present the market for early thinnings is weak and there is a tendency not to thin. An additional outlet for the product of early thinnings could be fuel' (Mitchell & Matthews, 1981). Mitchell & Matthews (1981) estimate that whole-tree harvesting of early thinnings could yield 300 kt DM (6 PJ) at present (1980) rising to 600 kt (12 PJ) by 2000 (equivalent to 0.1% and 0.2% 1981 UK primary energy consumption).

Table 3.6
Annual production and availability of horticultural wastes, UK, 1976[1]

Crop	Production		
	Total	Available[2]	
	kt DM	kt DM	PJ
Orchard (grubbings)	23	21.0	0.37[3]
Orchard (prunings)	14	14.0	0.25[3]
Raspberry	9	9.0	0.14[3]
Hop	10	10.0	0.10[4]
Tomatoes (glasshouse)	na	4.2	0.04[4]
Carnations (glasshouse)	na	0.7	0.01[4]
Total		58.9	0.91
% 1981 UK primary energy consumption[5]			0.01

1 Glasshouse crops — year not specified in source.

2 Available production refers to material of no value other than as a fuel.

3 Adjusted to allow for latent heat required to remove water when residue is burnt.

4 Assuming net methane yield *via* anaerobic digestion of 9.75 GJ per tonne DM.

5 8 369 PJ.

Sources: Adapted from Spedding *et al* (1979a); Larkin *et al* (1981)

Table 3.7

Annual production and availability of animal wastes, UK, 1976

Animal	Production		
	Total[1]	Available[2]	
	kt DM	kt DM	PJ
Dairy cattle	2 653	2 653	15.97[3]
Beef cattle	1 984	1 984	11.94[3]
Poultry	1 542	1 522	13.45[4]
Pigs	984	977	9.56[5]
Total	7 163	7 136	50.92
% 1981 UK primary energy consumption[6]			0.61

1 Total production of housed cattle and all poultry and pigs.

2 Available production is of housed farm animals only.

3 Assuming net methane yield *via* anaerobic digestion of 6.02 GJ per tonne DM.

4 Assuming net methane yield *via* anaerobic digestion of 8.84 GJ per tonne DM.

5 Assuming net methane yield *via* anaerobic digestion of 9.78 GJ per tonne DM.

6 8 369 PJ.

Source: Adapted from Larkin *et al* (1981).

Farm woodland

Many small areas of unexploited woodland on farms may be profitably managed towards the production of timber, firewood and other commodities. This would also lead to benefits to environmental conservation, including historic, landscape, wildlife and recreational interests, and to rural economy and employment (DART, 1983).

According to MAFF (1980) there were 190.9 kha of 'woodland in agricultural holdings' in England and Wales in 1979. Assuming an average output of 4 t DM per ha per year of fuelwood and energy content of 20 GJ per tonne DM, this amounts to 15.3 PJ gross energy in fuelwood or 0.18% 1981 UK primary energy. This estimate only serves as an approximate indication of the magnitude of the resources as:

(i) the quoted area does not include all the woodland which may be available on farms. DART (1983) suggest that the area of unmanaged woodland, mostly coppice and scrub, in England and Wales, is nearer 500 kha;

(ii) the estimate is restricted to England and Wales.

At the same time much of the existing woodland may be difficult to exploit. However, where the woodland is small single stem trees or coppice, easily accessible and close to a ready market, it may provide a significant addition to the farmer's income.

Management of small areas of farm woodland for fuel production would be compatible with recreational uses, such as pheasant rearing. It may even be possible to link coppiced woodland management for fuelwood with venison production from Chinese water deer *(Hydropotes inermis)*, a system currently under investigation by the Department of Agriculture and Horticulture and the Department of Zoology, of the University of Reading.

Table 3.8

Annual production and availability of wood residues, UK, 1977 (and forecast for 2000).

Source	Production							
	Total		Recoverable		Available for fuel[1]			
	kt DM		kt DM		kt DM		PJ	
Year	1977	2000	1977	2000	1977	2000	1977	2000
Forestry residues								
Roots & stumps	700	1 600	140	320	140	320	2.8	6.4
Tops & branches	700	1 600	231	1 200	231	1 200	4.6	24.0
Processing residues	na	na	1 472	2 400	712	920	14.24	18.4
Total	N/A	N/A	1 843	3 920	1 083	2 440	21.64	48.8
% 1981 UK primary energy consumption[2]							0.26	0.58

1 Material not used or likely to be used for various non-fuel uses.

2 8 369 PJ.

Sources: Adapted from Ader & Buck (1979); Mitchell & Matthews (1981).

Rural natural vegetation

Harvesting natural vegetation for biofuel production, in common with the opportunities already discussed, is a means of exploiting an unused resource; hence, competition with other land uses is minimal, and energy feedstock costs relatively low. The potential of this opportunity has been studied for a number of years at the Institute of Terrestrial Ecology, Merlewood Research Station, Cumbria.

According to Lawson & Callaghan (in press), 8.22 Mha (34.5%) of UK land is covered by vegetation which has not been purposefully grown for agriculture and forestry; most of this is in the hills of northern Britain and much will appear under 'rough grazing' in other classifications. A breakdown of the composition of this land area and the estimates of the productivity and total production of the different vegetation types appear in Table 3.9

Table 3.9

Extent and estimated production of rural natural vegetation in the UK

	Extent kha	Above-ground productivity $tDM\,ha^{-1}y^{-1}$	Total productivity $tDM\,ha^{-1}y^{-1}$	Above-ground production $ktDM\,y^{-1}$	Total production $ktDM\,y^{-1}$
Heather (GB)	1 490.0	3.0	6.3	4 470	9 387
Bent grass (GB)	1 080.0	3.7	7.3	3 996	7 884
Purple moor grass (GB)	700.0	5.0	8.0	3 500	5 600
Cotton grass/ Deer grass (GB)	530.0	2.5	6.3	1 325	3 339
Fescue (GB)	410.0	3.7	7.3	1 517	2 993
Meadow grass (GB)	380.0	8.8	13.2	3 334	5 016
Cocksfoot (GB)	330.0	9.1	13.6	3 003	4 488
Bracken (GB)	320.0	9.8	14.0	3 136	4 480
Mat grass (GB)	300.0	1.9	2.9	570	1 260
Blaeberry (GB)	140.0	2.1	4.2	294	588
Other rough grazing (GB)	447.0	2.8	4.1	1 252	1 833
Total rough grazing in GB	6 127.0	4.2	7.4	26 397	46 868
Total rough grazing in N. Ireland	210.0	4.2	7.4	882	1 554
Scrub woodland (UK)	298.3	6.8	9.1	2 028	2 714
Ungrazed mountain (GB)	745.2	2.4	4.8	1 788	3 577
Foreshore (GB)	377.3	5.0	10.0	1 887	3 773
Hedges (GB)	181.0	4.8	7.2	869	1 303
Nettle (GB)	331.0	9.7	14.5	321	480
Gorse (EW)	26.4	16.8	18.7	444	494
Saltmarsh (GB)	19.4	12.0	21.4	233	415
Willow herb (GB)	14.4	11.0	16.5	158	238
Marsh, fen & carr (EW)	14.1	10.8	16.2	152	228
Meadow sweet (GB)	11.0	6.5	9.8	72	108
Butterbur (GB)	3.3	8.2	12.5	27	41
Other misc (UK)	462.8	7.0	10.5	3 240	4 859
Total	8 523.3	4.5	7.8	38 498	66 652

Note: Data from various years from 1958–1978

Source: Adapted from Lawson & Callaghan (in press)

This vegetation represents a resource of about 643 PJ per year in gross energy terms, but much of the vegetation may be unharvestable due to slope, boulders etc, some may already be used as rough grazing for animal production and productivities are unlikely to be maintained with repeated cropping.

Of these areas of natural vegetation, Lawson *et al* (1980) consider that heather moors and bracken fells provide the most suitable areas available immediately for cropping. Particular attention has been given to bracken, which could be harvested dry for direct combustion or gasification to methanol, or fresh for anaerobic digestion (Callaghan *et al,* 1982). An assessment of the potential bio-fuel output of some types of natural vegetation appears in Table 3.10. These figures assume a high exploitation of bracken and heather and harvesting of only a small proportion of other forms of vegetation.

Table 3.10

Potential annual biofuel output of rural natural vegetation in the UK

	Extent	Harvestable production	
Vegetation	kha	kt y^{-1}	PJ y^{-1}
Bracken	320	2 048[2]	25.80[7]
Heather	1 490	1 192[3]	15.02[7]
Rough grass	3 940	1 088[4]	10.61[8]
Scrub woodland	107[1]	214[5]	4.28[9]
Other	2 359	236[6]	2.97[7]
Total	82 167	4 778	58.68
% 1981 UK primary energy production[10]			0.70

1 Net of 'scrub woodland' in Table 3.9 and 'farm woodland' of previous section.

2 Assuming sustainable harvestable production of 8t DM per ha per year and 80% area harvestable.

3 Assuming 20t DM per ha in 20 years and 80% harvestable.

4 Assuming 60% of production from Table 3.9 sustainable and 10% harvestable.

5 Assuming 2t DM per ha per y

6 Assuming 1t DM per ha per y and 10% harvestable.

7 Assuming net fuel yield of 12.6 GJ per t DM.

8 Assuming methane production via anaerobic digestion of 9.75 GJ per t DM.

9 Assuming 20 GJ per t DM.

10 8 369 PJ.

Notes: Data from various years.

Sources: Adapted from Callaghan *et al* (1982); Lawson & Callaghan (in press).

Urban vegetation

Various urban areas, including sports grounds, parks, domestic lawns and gardens, street and motorway verges and railway embankments, produce vegetation which is unused. Some of this vegetation is cut already and it should be possible to alter management methods to include economic collection of a proportion of the biomass. Lawson & Callaghan (in press) estimate the total dry matter production of urban vegetation to be 8444 kt DM per year and consider 20% of this to be realistically collectable. This is equivalent in energy terms to 16.5 PJ per year as methane via anaerobic digestion or 0.2% UK 1981 primary energy consumption.

Aquatic and marine vegetation

Waterweeds

About 15–40 kt DM of waterweeds (ie floating-leaved and submerged macrophytes) grow, each year, in UK rivers, canals and larger drainage channels of which 10–20 kt are harvested (Newman, 1979). Assuming anaerobic digestion to methane at 9.75 GJ per tonne the harvested proportion represents an energy resource of 0.09–0.20 PJ per year.

Seaweed

Newman (1979) has estimated the total UK seaweed resources to be about 1.3–1.9 Mt organic matter. Harvested areas should regrow in four years making an annual production in the range 0.3–0.5 Mt organic matter, equivalent to a potential energy production (as methane) of 2.9–4.9 PJ per year. Inland waterweeds and seaweed represent together a biofuel resource equivalent to 0.04–0.06% 1981 UK primary energy consumption.

Multipurpose crops

Unlike the opportunities considered so far, which are all based on existing resources, the use of multipurpose crops involves a system change designed to increase the amount of unused material and divert it to fuel production. However, prolonged use of crop residues (or other byproducts of a particular process) for fuel also involves a system change and the residue or byproduct should not be regarded as a waste, but as a useful (even if secondary) output of the system.

The use of cereal straw for fuel and grain for food is an example of a dual-purpose crop with a fuel component. It may also be possible to replace current crop species with species that produce food, feed and fuel components which can be easily separated. The growing of plantations of multipurpose crops that can be industrially fractionated into the basic components of human food, fuels and other raw materials, may also be possible, but this would require major changes in technology, land use and human diet.

Hence, the system changes possible are:

(i) use of all components or fractions of existing crops;

(ii) growing species specifically for a number of components;

(iii) use of technology to separate out all the desired components from crops grown simply for high biomass yield.

The most promising dual-purpose crops for the provision of food/feed and fuel components are oilseed rape, Jerusalem artichokes, fodder beet and sugar beet (Spedding *et al*, 1979b). Oilseed rape provides: vegetable oil which is currently used in foods and in industry and which can be used as a fuel to mix with or replace diesel oil (Bühner & Kögl, 1981; American Society of Agricultural Engineers, 1982); a protein-rich meal used as animal feed; and straw suitable for animal feed or bedding or use as a fuel. This is illustrated in Figure 3.2.

Fodder beet and sugar beet produce tops which can be fed to animals or be anaerobically digested to yield methane. The roots can also provide feedstocks for anaerobic digestion to methane or for ethanol production in addition to their traditional use as animal feed or in sugar manufacture. Jerusalem artichoke *(Helianthus tuberosus)* tubers provide a suitable feedstock for ethanol or methane production and can be used as animal feed or human food, and the tops can be harvested fresh for animal feed or as an anaerobic digestion feedstock, or dry as straw for direct combustion or thermal processing. The residue from ethanol production may also be suitable as an animal feed or even a feedstock for anaerobic digestion (see Figure 3.3).

Assessment of the potential UK output of fuel from dual-purpose crops is difficult owing to the problems associated with the calculation of the area of available land. An assessment of potential rape oil and straw output at different levels of land availability appears in Table 3.11. (This estimate is based on 1977 crop areas.) It should be noted that although rape oil energy output and straw energy output have been summed to provide a total estimate of potential energy yield, rape oil is of appreciably greater value than straw as it provides a liquid fuel with little further processing; conversion of straw to a liquid fuel would involve energy losses and processing costs. Replacement of crops grown for animal feed with oilseed rape would provide 93.2 PJ of energy in oil and straw equivalent to 1.11% of 1981 UK primary energy consumption.

Table 3.12 presents an estimate of the potential UK output from the dual-purpose root crops, sugar beet and Jerusalem artichoke based on their replacing 0.5 Mha of animal feed crops. The gross energy as fermentable carbohydrate is not equivalent to energy yield as ethanol and further reductions can be expected. Use of either roots or tops to produce methane excludes any other use although the residue from anaerobic digestion is still of value as a fertiliser; ethanol production from roots also provides an animal feed fraction.

Figure 3.2
Oilseed rape as a multipurpose crop

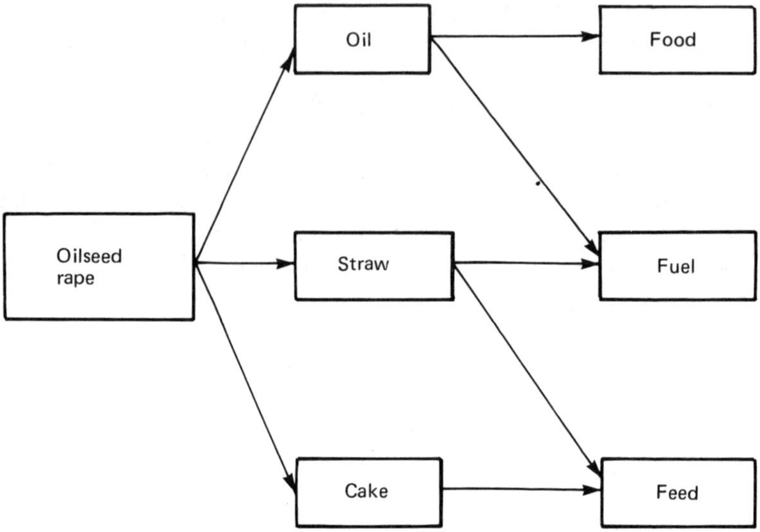

Figure 3.3
Jerusalem artichoke as a multipurpose crop

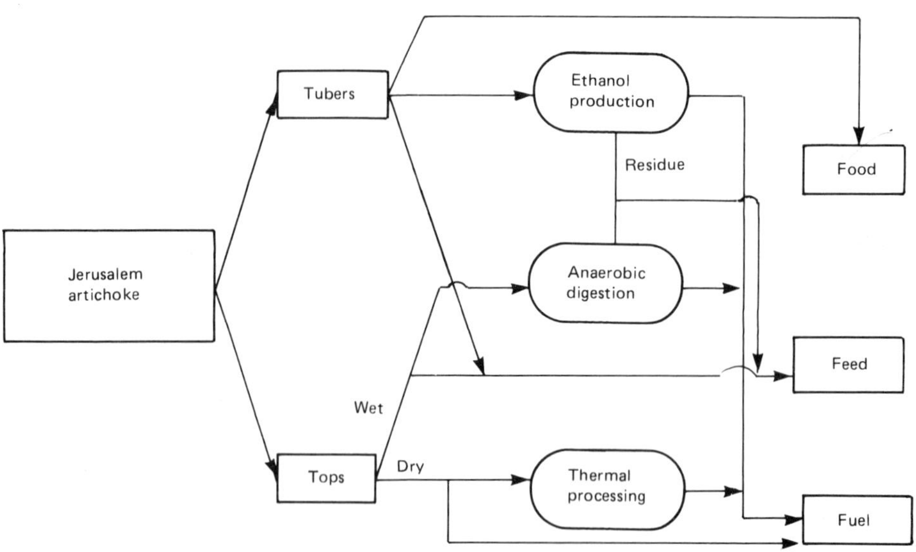

Table 3.11

Potential annual output of energy from oilseed rape under different levels of land availability, UK, 1977

Area definition	Area Mha	Total[5] oil output kt	Oil[6] energy output PJ	% 1981 UK[9] primary energy consumption	Total[7] straw output Mt	Straw[8] energy output PJ	% 1981 UK[9] primary energy consumption	Total (straw & oil) energy output PJ	% 1981 UK[9] primary energy consumption
Current oilseed rape area	0.048	43.3	1.69	0.02	0.403	7.25	0.09	8.94	0.11
Current tillage[1] break area	0.3	270.8	10.56	0.13	2.52	45.36	0.54	55.92	0.67
Current animal feed area	0.5	451.3	17.60	0.21	4.20	75.60	0.90	93.2	1.11
20% current arable area	1.3	1 173.3	45.76	0.55	10.92	196.56	2.35	242.32	2.9
Current break[2] area	2.3	2 075.8	80.96	0.97	19.32	347.76	4.16	428.72	5.13
Current arable[3] area	6.7	6 046.8	235.83	2.83	56.28	1 013.04	12.10	1 248.87	14.92
Overall[4] suitable area	11.32	10 216.3	398.44	4.76	95.09	1 711.62	20.45	2 110.06	25.21

1 Total tillage area not devoted to cereals, sugar beet, potatoes, vegetables, horticulture or orchards.

2 Total tillage break area plus temporary grass.

3 Total tillage plus temporary grass.

4 Calculated using the ITE Land Classification System.

5 Assumes oil yield of 0.9025 t per ha (57% mean winter rape yield plus 43% mean spring rape yields).

6 Energy content of rape oil 39 GJ per t.

7 Assumes straw yield of 8.4 t per ha.

8 Assumes energy content of straw 18 GJ per t.

9 8 369 PJ.

Source: Adapted from Spedding *et al* (1979b).

Table 3.12

Potential annual output of multipurpose root crops, UK, 1977

Crop		Sugar beet			Jerusalem artichoke		
Fraction		Output ha^{-1}	UK output[1] ($\times 10^6$)	% 1981 UK primary energy consumption[2]	Output ha^{-1}	UK output[1] ($\times 10^6$)	% 1981 UK primary energy consumption[2]
Roots	Total (tDM)	9.9	4.9	N/A	4.7	2.4	N/A
	Energy as[3] methane (GJ)	96.5	48.3	0.58	45.8	22.9	0.27
	Energy in fermentable carbohydrate (GJ)	111.0	55.5	0.66	49	24.5	0.29
	Fermentation byproducts (tDM)	3.4	1.7	N/A	1.8	0.9	N/A
	Protein in byproducts (t)	0.5	0.25	N/A	0.4	0.2	N/A
Tops	Total (tDM)	5.8	2.9	N/A	10.0	5.0	N/A
	Energy as methane[3] (GJ)	56.6	28.3	0.34	97.5	48.8	0.58
	Protein in tops (t)	0.7	0.35	N/A	1.1	0.55	
Roots & tops	Total protein (t)	1.2	0.6	N/A	1.5	0.75	N/A

1 Assuming crop replaces 1977 area of annual feed crops (0.5 Mha).

2 8 369 PJ.

3 Assuming 9.75 GJ per tonne

Source: Adapted from Spedding *et al* (1979b)

INCREASING PRIMARY PRODUCTION

An increase in total primary production would make more biomass available for biofuel production unless the demand for other products took precedence. In practice, an increase in primary production towards biofuels would either involve fuel crops being grown on land currently unused or under-used, or land being released from the provision of other products by increases in their productivity. Strategies for increasing primary production towards fuel production include:

(i) Using unused land for fuel crops:
  (a) derelict land;
  (b) other unused land or under-used land;
  (c) via catch cropping;
  (d) via mixed cropping.

(ii) Releasing land for fuel crops by increasing crop yields.

Derelict land

The areas of derelict, despoiled and degraded land in GB suitable for growing energy crops has been estimated as some 0.25 Mha, capable of yielding 1 Mt dry matter per year of coppiced woodland (Dennington & Chadwick, 1979). At 20 GJ per tonne this amounts to 20 PJ gross energy in wood, equivalent to 0.24% 1981 UK primary energy consumption. Dennington and Chadwick (1979, 1981) did not assess land in Northern Ireland. Derelict land arises mainly from the mineral working and processing industries, rail and military closures and industrial decline, but the above area includes other degraded sites in urban industrial areas including disused allotments, vacant sites, overgrown tips and spoil heaps and general waste land (Dennington & Chadwick, 1981).

Other unused or under-used land

There is inevitably some overlap between estimates for the areas of rural natural vegetation and urban vegetation, and estimates of the areas of unused land or derelict land available for fuel cropping. The existing production of the area of derelict, despoiled and degraded land in Britain will have been included in the estimate for total production of urban vegetation, as would some of the rural areas considered to be available for fuel crops have been included in the production for rural natural vegetation.

However, the recommendations of Dennington & Chadwick (1979, 1981) referred to above aim to increase the productivity of derelict land (largely 'urban') by cultivation and tree planting rather than to exploit the existing vegetation.

Similarly, Mitchell & Matthews (1981) estimate that 1.5 Mha of rural land are potentially available for wood energy plantations in the UK; 0.5 Mha comprises small areas of unused or under-used land such as field and stream margins, rough grazing and unmanaged scrub woodland in the lowlands, and 1 Mha unused or under-productive marginal agricultural land in the uplands. This land, if used for single stem forestry energy plantations could produce 10 Mt DM per year or in energy terms 200 PJ or 2.4% 1981 UK primary energy consumption.

The production of some of this land, using the existing vegetation, will already have been estimated in the farm woodland or rural natural vegetation categories and some of it is agriculturally productive, albeit at a very low level. Hence, this strategy involves a change in the use of the land and may in some cases involve a decrease in agricultural output (mostly animal products), if the land is considered under-used rather than unused. This is discussed in a later section of this Chapter.

Catch crops

Catch crops are grown between the harvest and sowing of main crops, and hence use land which is otherwise unused for part of the year. Traditionally, catch crops have been used to provide animal feed in the late autumn-winter period, but with the decline in mixed farming many areas of arable land lie fallow at this time of year. The introduction of catch crops into arable rotations to provide a biofuel feedstock is a strategy which:

(i) need not displace food production;
(ii) is relatively easily integrated into current cropping regimes;
(iii) would use machinery and other resources already available on the farm;
(iv) may use labour at an otherwise slack period;
(v) may provide fuel when demand is high.

The conversion process most suited to the wet, green feedstock usually provided by catch crops, would be anaerobic digestion to methane. In the longer term, development of the technology associated with the hydrolysis of cellulosic materials and the fermentation of the product may result in the 'ethanol route' becoming a more attractive option for the conversion of catch crop material. Additionally, starchy root crops which can be sown if the land becomes available in June provide a more suitable substrate for ethanol production, although the opportunity represents only a small area.

The potential for growing catch crops for fuel in the UK depends on productivity and the extent of implementation. Productivity is a function of crop yield and conversion efficiency; implementation will be influenced by land suitability, land availability, farm suitability, costs and the existence of a market for the fuel (Carruthers, 1982, 1983). Potential output has been estimated on the basis of an assessment of land availability based on current crop areas, experimental yield data and a conversion efficiency of 9.75 GJ methane per tonne of crop dry matter. This places output at 174 PJ per year or 2.1% of 1981 UK primary energy consumption.

The growing of catch crops for fuel has been studied at the University of Reading for several years and is reported elsewhere (Spedding *et al,* 1979b, 1980; Bather & Carruthers, 1980; Spedding & Carruthers, 1980; Carruthers, 1981a, 1981b, 1982, 1983).

Although catch crops are here considered as potential fuel crops they could also be employed more extensively to provide winter feed, hence increasing the overall productivity per unit area of land, and potentially making other land available for fuel crops (see section on crop yield increases).

Mixed cropping

While catch cropping represents intensification only in the time dimension, mixed cropping involves increased exploitation of space and, in some cases, time. In the majority of cases, mixtures of species, or of different varieties of the same species, show little advantage over pure stands, but some mixtures exhibit a yield increase over the yield of an equivalent area divided equally between the two species and, in some cases, a combined yield greater than the yield of an equal area of the best yielding component. In a small number of cases mixtures have proved to be disadvantageous (Trenbath, 1974).

Suggested advantages of mixtures include yield increases, stability over seasons, spread of production over growth period, less susceptibility to disease and lodging and improved quality. The biological basis for the yield increases observed in some mixtures is, primarily, their enhanced efficiency of use of light, nutrients and water (Willey, 1979).

Various types of mixed cropping can be practised including random or intimate mixtures (eg grass and legumes), row, strip and relay inter-cropping (Sanchez, 1976).

Although mixed cropping may be a means of increasing the biological and economic yield of agricultural land, there are also a number of disadvantages, particularly in relation to the mechanisation of mixed crops, and further research is needed. If mixed cropping is to be seen as an opportunity to grow a fuel crop, the fuel crop needs to be grown in combination with currently grown food or animal feed crops and must, if possible, have no deleterious effect on food or feed crop yield and be easily separated from the food or feed crop.

The introduction of a fuel crop into current crop populations will depend on the degree of spacing and ground cover of the current crop. The greatest potential lies in orchards and possibly peas, beans, roots and other row crops. These crops cover only a small proportion of the UK land area and overall potential is low, although such systems may make a substantial contribution on individual farms (Spedding *et al,* 1979b).

Crop yield increases

Crop yields have increased quite considerably since the war, largely as a result of new varieties, increased use of inorganic fertilisers and agro-chemicals for the control of weeds, pests and diseases (Dermott, 1978). In the years from 1947 to 1975 national average wheat yields increased by 84%, 50% being attributable to variety, and barley yields by 63%, 30% as a result of variety (Silvey, 1978). However, there is still quite a difference between average yields and theoretical maximum yields, yields in trials and yields on particular farms in specific years. Models suggest that yields of 12 t grain per ha of wheat and 90 t per ha of potatoes are possible (Dermott, 1978) and the latter may be an under-estimate (P M Harris, personal communication).

Further increases in crop yields therefore seem possible. The potential of mixtures and catch crops to achieve increases in overall productivity by employing more than one species has already been discussed. Crop yield is the result of the interaction between genotype and environment; hence, the yields of monocultures may be improved by manipulation of the genotype, by breeding and selection, and of the environment, by management, and by a careful matching of one to the other.

Breeding and selection, as shown above, have been important determinants of yield increases in the past. As criteria and resources change new varieties and species are needed. For example, many crop varieties have been bred to perform best at high inputs of inorganic fertilisers; if these become increasingly expensive, growers may be looking for varieties which will perform well with lower inputs. Similarly, if the product demanded changes (as with transfer from a food to a fuel crop) then new genetic material is needed. Breeding and selection have, therefore, the potential to sustain yields under changing circumstances and, in some cases, to bring about improvements.

The use of the genetic material most suited to a specific, sometimes very localised, environment may bring about significant improvements in crop yield. Varieties are often produced independent of use and local varieties may be of considerable value, particularly where high levels of management inputs cannot be sustained. Wareing & Allen (1977) indicate that a geographical redistribution of agricultural systems within the UK, and potentially the EC, towards optimum bioclimatic regions would increase yield and could, thus, release resources for biofuel production.

In many cases, higher yields may be achieved by more precise manipulation of the environment (ie management) and greater attention to detail. Important management factors include fertiliser applications, both quantity and timing, control of pests and diseases, sowing date and irrigation. Timeliness is particularly significant, and a growing awareness of this may promote further yield increases.

Increased use of irrigation may also be an important determinant of yield in the future, in the same way that varieties and agro-chemicals have been in the past (P M Harris, personal communication).

A consideration of ruminant feed production provides a specific example of this strategy.

Ruminant feed production

According to Wilkins *et al* (1981) the UK currently devotes 7.1 Mha of grass (excluding rough grazings) and the equivalent of 2.2 Mha of cereal land to ruminant feed production, equivalent in total to about 44% UK agricultural and forestry land. Wilkins *et al* (1981) estimate that, using improved management systems, the current UK output (consumption) of ruminant products could be provided by 0.87 (1.2) Mha of cereals and either 3.01 (4.09) Mha of grass with optimum levels of N fertiliser, or 3.87 (5.19) Mha of grass and white clover using a legume-based forage production system. The latter system may be more desirable than the former as the financial and energy costs of nitrogen fertiliser may prohibit the required increase of some 47%.

One effect of these improvements would be to release land for other purposes. Much of the land released would be suitable for other crops and all would be suitable for some kind of forestry. Hence, biofuel production, either *via* arable fuel crops or wood energy plantations, would provide a suitable enterprise for this land. Table 3.13 provides an estimate of the potential effect of the scenarios for ruminant feed production of Wilkins *et al* (1981) on biofuel production, using wood energy plantations as an example.

Integrated production systems

There is some evidence that integrating forestry, to produce timber and/or fuel-wood, with crop and animal production (ie agroforestry) is more efficient, in terms of land use, than producing the commodities in separate enterprises. Similarly systems designed to integrate a number of enterprises and use the by-products of one enterprise as the inputs of another may prove more efficient in terms of both energy and land use. Small-scale labour-intensive systems producing a variety of foods and other products may also use less land to produce the same outputs than more conventional systems. The possibilities require further investigation, but, if successful, would reduce the amount of land required for food crops and make more land available for fuel production.

Table 3.13

Potential wood energy output of land released by the ruminant feed production scenarios of Wilkins *et al* (1981).

Scenario	Area released Mha	Potential wood energy output[1] PJ	% UK 1981 primary energy consumption[2]
To produce current UK output			
Grass with N fertiliser	5.42	1 084	12.95
Grass with white clover	4.56	912	10.90
To produce current UK consumption			
Grass with N fertiliser	4.01	802	9.58
Grass and white clover	2.91	582	6.95

1 Assuming yield of 10 t DM per ha per y

2 8 369 PJ.

Source: Wilkins *et al* (1981).

DECREASING INPUT TO SECONDARY PRODUCTION

Secondary production involves considerable energy losses. A decrease in the amount of primary production going to secondary production could be achieved without an overall increase in the quantity (in energy terms) of food and other commodities, although it may involve a change in quality. A decrease in the inputs to secondary production would make primary production, or the land which provides it, available for fuel production. Such a decrease in the input to secondary production could be achieved by:

(i) increasing the efficiency of animal production in terms of land;

(ii) dietary change toward vegetable products.

Increasing the efficiency of animal production

Improvements in the efficiency of animal production in terms of feed conversion could be achieved by an avoidance of 'compositional overproduction'. This would involve producing animal products according to compositional specifications, thus reducing the production of unwanted byproducts and, hence, matching the supply, in terms of product quality, more closely to the demand.

An example of this is provided by milk production in the UK. CAS, in a report to the House of Commons Agriculture Committee (1980) on the dairy

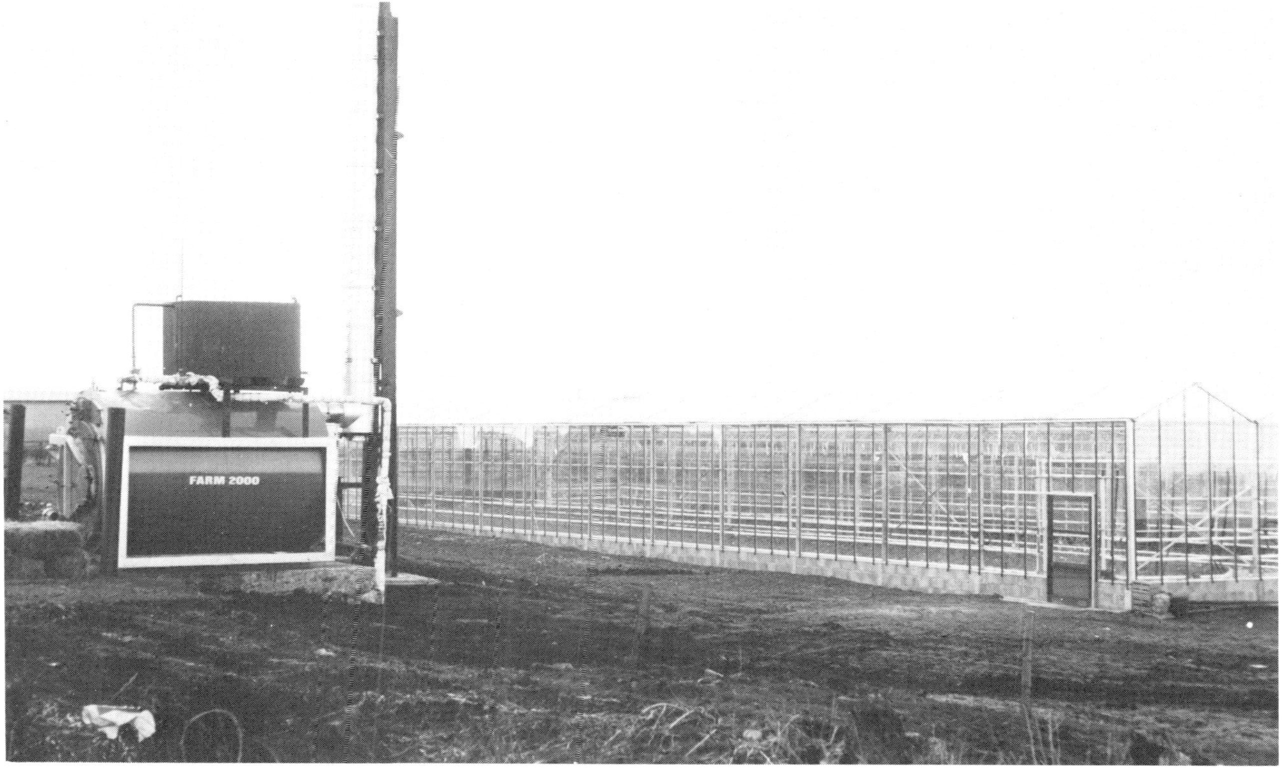

Direct combustion of dry feedstocks such as straw and wood can be used for space heating. This boiler uses large round straw bales to heat a 500 m^2 glasshouse at Cleveland, County Durham. Photograph courtesy of Farm 2000 Ltd.

Anaerobic digestion may be used to produce methane-rich biogas from wet feedstocks such as animal wastes and green crop residues. This digester operates on wastes from a pig unit at Walnut Tree Farm, Chediston, Suffolk. Photograph courtesy of Hamworthy Engineering Ltd.

Biogas from the anaerobic digestion of wastes from a 340-cow dairy unit is used to generate electricity for on-farm use at Bore Place Farm, Edenbridge, Kent.

Bracken and farm woodland in the uplands represent a short-term opportunity for biofuel production. Lowick Common, Cumbria. Photograph courtesy of T V Callaghan.

Catch crops for fuel production could be integrated into existing arable systems. This crop of fodder radish, grown in a field experiment at the University of Reading, yielded 5t DM per ha 12 weeks after sowing.

The naturalised alien, Japanese knotweed, appears to be capable of yields of up to 20t DM per ha per year. This stand was found in Rochdale Road Cemetery, Manchester. Photograph courtesy of T V Callaghan.

Coppicing, the cutting of established trees to regenerate multiple stems, is a traditional form of woodland management, and may be used to improve derelict woodland. The photograph shows detail of a coppiced poplar 'stool' in a small-scale woodland management project at the University of Reading.

Short rotation coppice yielding up to 20t DM per ha per year, could be used for large-scale wood energy plantations. A prototype harvester has been developed at the Loughry College of Agriculture and Food Technology. Photograph courtesy of C P Mitchell.

industry, considered that the composition of average milk has been out of line with the main market requirements over a long period (since before 1951); this compositional overproduction is restricted to lactose and protein — the components of the byproducts skimmed milk and whey. Changes in milk composition could be effected in the long term using genetic selection and changes in the feed. The conclusion was that, among other benefits, the production of milk to compositional specifications was biologically and technologically feasible, would use farm resources and feed-stuffs to better advantage, and would reduce the production of unwanted byproducts.

There are a number of other ways in which animal production efficiency might be improved:

(i) by smaller livestock units integrated with arable, horticultural or food processing enterprises and using wastes from these enterprises. It has been estimated that a particular horticulture station growing 24 ha of vegetables is capable of supporting 12 head of beef cattle (M A Wagner, personal communication);

(ii) by the use of animal species with a better exploitation of the breeding population. Animals with a large litter size, such as rabbits and Chinese water deer, divert less food energy, on a population basis, to maintenance of breeding stock, than species with small litters such as sheep and cattle (Large, 1973);

(iii) by the use of animal species with a higher biological feed conversion efficiency, for example, poikilotherms such as fish and invertebrates (Spedding *et al,* 1981).

Animals provide an alternative to anaerobic digestion as an outlet for wet crop and vegetable wastes, and this may be a more feasible route, particularly if land is released for the growing of fuel crops which provide liquid fuels. Many vegetable processing factories produce considerable quantities of waste and this is usually dumped, often at some expense; use as an animal feed provides an alternative to this (Fairbairn *et al,* 1971). Integration of animal production with growing crops on a small scale may also prove more efficient, both energetically and in terms of land use.

Although these suggestions would not mean a decrease in overall consumption of animal products in the nation's diet, they would involve a change in the quality of these products, and a move from traditional meats to more 'exotic' foods such as venison, rabbit, carp, trout or perhaps, even snails.

Dietary changes towards vegetable products

The proportion of agricultural land used to provide food for direct human consumption compared with that devoted to animal feed is indicated in Table 3.14. According to Leach (1976) about 91% of UK agricultural land is rough grazing, temporary or permanent pasture or crops grown to feed animals while 9% provides vegetable foods for human consumption. At the same time UK agriculture provides almost equal quantities of dietary energy in vegetable foods as in animal products. This indicates that, on average, the vegetable system is over ten times as efficient as the animal system in terms of its use of land resources to provide dietary energy, and suggests that the replacement of animal products in the diet with foods derived directly from primary production could, therefore, release appreciable quantities of land for fuel production.

Table 3.14

Proportions of agricultural land used to provide food or feed, UK, 1971/72

	Food %	Feed %
Cereals	25	75
Other field crops	67	33
Grass and lucerne	0	100
Total crops and grass	14	86
Rough grazing	0	100
Total all land	9	91

Source: Adapted from Leach (1976)

An approximate idea of the potential of this strategy is provided by considering the replacement of animal feed crops and grass with wood energy plantations, on the area of cultivable land used to provide animal feed. According to Leach (1976) this area amounted to 86% of the total area of arable crops and grassland (Table 3.14) in 1971/72. The remaining 14% provided food from crops. Assuming that a further 14% of the total area would more than adequately replace the dietary energy of the displaced animal products a net 72% of arable land and grassland would be released. Applying this to the 1976 land areas (Table 3.1) an area of 8.69 Mha is made available, capable of providing 1738 PJ per year of fuelwood (assuming 10 t DM per ha per year and 20 GJ per tonne) or 21% 1981 UK primary energy consumption.

Rough grazing land is not suitable for arable crops, but some areas could be used for wood energy plantations, thus increasing the potential fuel output of this strategy, and remaining areas could continue to provide animal products. Additionally livestock could be fed on crop residues, food processing wastes and wastes from biofuel production processes.

Adoption of this strategy would mean an appreciable reduction in consumption of animal products, but it would not be necessary or desirable to eliminate animal production altogether.

DECREASING OUTPUTS

An overall decrease in the indigenous production of food and other products would release resources for biofuel production or land for fuel crops. The simplest situation where this would occur is where a transfer of land use from the current enterprise to biofuel production increased the profitability of that land.

An example of this is provided by Mitchell *et al* (1983) who investigated the availability of land for the production of wood for energy in GB. Using the ITE Land Classification System as a sampling system and a data base, the financial returns from the existing land uses were compared with those projected for:

(i) conventional forestry;

(ii) modified conventional forestry, producing both timber and wood for energy;

(iii) single stem energy forestry;

(iv) coppice energy forestry.

In addition the 'effects of other demands that society makes on the land were examined, including the presence of National Parks, Nature Reserves and the impact that changes to forestry might have on landscape and wildlife conservation'. These effects were termed 'social and institutional constraints'.

The study examined a wide range of energy and timber prices, agricultural and forestry costs and returns, and discount rates, but only the results of an 'exploratory case' are reported in Mitchell *et al* (1983). This assumed:

(i) wood produced for conversion to pipeline gas;

(ii) increasing energy prices;

(iii) constant timber and agricultural costs and revenues;

(iv) a 5% discount rate.

A summary of the results of this study appears in Table 3.15. Forestry never approached the financial performance of any crops and conventional forestry was never able to compete financially with current land uses or the other forestry systems. The essential conclusion, however, is that between 10% and 26% GB agricultural land, depending on social and institutional considerations, could give a better financial return with forestry than with the current use, and that this land is capable of providing from 320 PJ to 760 PJ of wood for fuel and from 11.0 Mm^3 to 28.0 Mm^3 of timber.

Most of this land would be gained from beef and sheep production. Mitchell *et al* (1983) did not report on the quantities of beef, sheep and other products which would be lost and the consequences of such losses if the strategy were to be adopted. Linking this strategy to the improvements in ruminant production already referred to, may significantly increase biofuel production with a minimal impact on the output of ruminant products.

Table 3.15

Land available for, and fuelwood and timber output from, different forest systems before (and after) allowing for social and institutional constraints, GB, 1977

	Area	Wood for fuel			Timber
Forest system	Mha	MtDM	PJ[1]	%[2]	Mm^3
Modified conventional forestry	2.9 (1.1)	16.0 (7.0)	320 (140)	3.8 (1.7)	28.0 (11.0)
Single stem	1.2 (0.4)	14.0 (4.0)	280 (80)	3.3 (1.0)	N/A N/A
Coppice	0.7 (0.3)	8.0 (5.0)	160 (100)	1.9 (1.2)	N/A N/A
Total	4.6 (1.8)	38.0 (16.0)	760 (320)	9.1 (3.8)	28.0 (11.0)

1 Assuming 20 GJ per t.

2 % 1981 UK primary energy consumption (8 369 PJ).

Source: Adapted from Mitchell *et al* (1983).

INCREASING IMPORTS

Although land may be transferred to biofuel production at the expense of indigenous food production if it is profitable, it is unlikely that national food consumption will decrease dramatically and, therefore, the loss would need to be made up by food imports.

Bioclimatic rationalisation in the EC

Increased reliance on food imports from the EC, and an accompanying increase in the opportunity for biofuel production within the UK, may also occur if a bioclimatic rationalisation of EC agriculture were to be implemented. This would involve the geographical redistribution of agricultural systems within the EC towards the optimum bioclimatic regions already referred to (Wareing & Allen,

1977). Individual regions would produce the crops best suited to their environments. The UK is best suited to the growing of grass and woodland, and much of the latter might be directed towards fuel production.

Economic rationalisation in the EC

Similarly, an economic rationalisation within the EC may increase UK reliance on food imports and indigenous production of biofuels. There are distinct regional disparities of efficiency of agricultural production (in terms of resource use) within the EC for different commodities (eg CAS, 1980b). If agriculture in the different States of the EC were better integrated than at present, and agricultural production were rationalised towards the more economically efficient areas for each commodity, then land could be released for biofuel production. The EC is nearly self-sufficient in most major food commodities and has a surplus in several. The opportunities for biofuel production within the EC would therefore seem greater than for the UK alone. The conversion of surplus commodities to fuel as is the case, for example in the US gasohol programme, would serve merely to entrench the existing situation of overproduction and would, therefore, seem unsuitable as a long-term solution to this problem. There have been several proposals to rationalise EC agriculture to avoid recurrent surpluses and some of these, notably the Mansholt and Vedel plans (CEC, 1968; Ministère de l'agriculture, Paris, 1969) suggested withdrawal of land from agricultural use. Such land, some of which might be in the UK, could be used for biofuel production.

OVERALL POTENTIAL

Although the overall potential of the strategies above can be estimated by summing individual output estimates, the results of doing so need to be treated with some caution as a number of the strategies represent alternative routes for the same resources. Additionally the quality of the energy produced varies with the strategy and a gross summation gives no indication of the type or value of fuels produced.

Very approximately, the exploitation of wastes, existing natural vegetation, growing catch crops and growing trees on derelict and other unproductive land could provide up to an equivalent of 14.5% 1981 UK primary energy consumption, mostly as combustible material such as wood and straw, and as methane.

Changes in the efficiency of ruminant feed production or dietary changes towards vegetable products could, in addition, provide between 7% and 21% UK primary energy consumption (depending on degree of change), primarily as wood. About 1.8 Mha of UK land could be profitably transferred to wood energy plantations or modified conventional forestry now and this figure rises to 4.6 Mha if social and institutional constraints are removed. These areas are capable of providing respectively 3.8 and 9.1% 1981 UK primary energy consumption as fuelwood. The factors affecting the realisation of this potential are discussed in Chapter 5.

4 Choice of crop for fuel production

A number of strategies described in the preceding chapter are based on the exploitation of existing biomass resources, either waste materials and byproducts or stands of natural or semi-natural vegetation. In these cases no choice regarding fuel crop species and variety needs to be made and the nature of the feedstock available will strongly determine the conversion process and resulting fuel. Other strategies, however, involve the utilisation of land opportunities, or the release of land from other uses, for the cultivation of crops partially (eg multipurpose crops, modified conventional forestry) or totally devoted to fuel production. These strategies therefore present varying degrees of choice as to the species and varieties grown. This chapter discusses the basis for choosing crops for growing on land released for fuel production and reviews the plant species available.

CHOICE OF FUEL CROP

In general, factors related to the land will interact with the type of fuel required, to determine the fuel crop system adopted. The relative influence of land quality and fuel demanded on the type of crop grown and the conversion process employed is strongly determined by economic factors.

One view of this effect, developed by Callaghan *et al* (1978) is presented in Figure 4.1. The two major constraints on the nature and degree of adoption of biofuel production systems are seen as the availability of land, as determined by land costs and the value of its present product, and the costs of biofuels in comparison with traditional fuels.

Figure 4.1
Diagrammatic representation of the biofuel production system

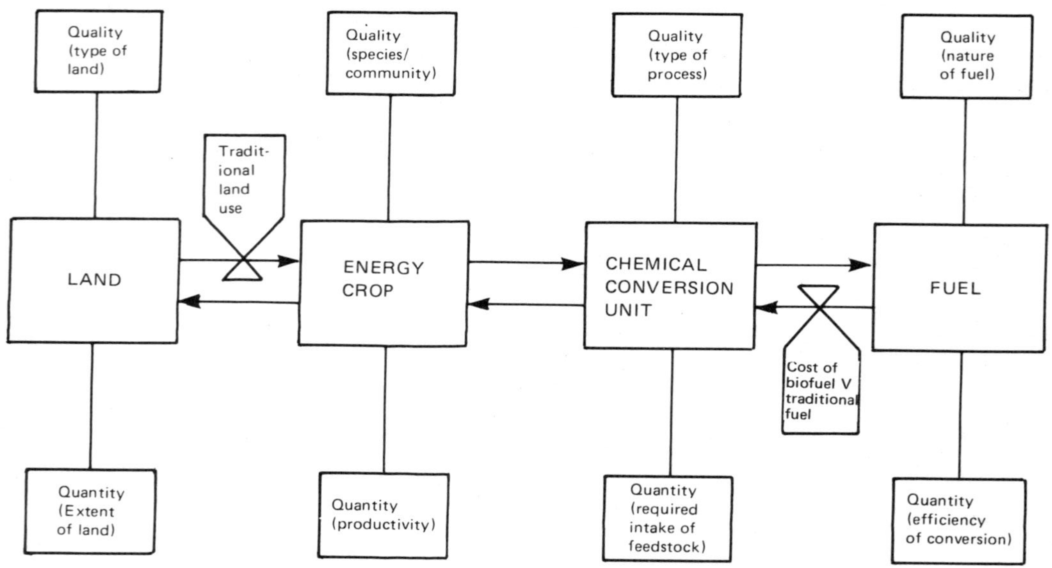

Source: Reproduced from Callaghan *et al* (1978)

In the short term the type of land and the quantity available for energy crops will be constrained by the traditional land use. Biofuel production will therefore be based largely on unexploited material (wastes and natural vegetation) and fuel crops are likely to be restricted to areas of derelict and marginal land. In this case the quality of the land will determine the type of fuel crop and, hence, the conversion process for the fuel.

However, a situation can be envisaged, in the longer term, where traditional fuel prices are such that it is economically feasible to replace a particular fuel with one obtained from plant material. In this case an energy crop may be more profitable to the farmer than the traditional crop and the process operates in reverse, the type of fuel required determining the fuel crop grown.

Land factors

Factors related to the land made available for fuel cropping which will influence crop choice include: environmental factors; social and organisational constraints; and duration of availability.

Environmental constraints include edaphic factors, such as soil type and soil fertility, and climatic factors related to altitude, latitude and location and other factors such as slope, accessibility etc. For example, in relation to wood energy plantations, high altitude locations would only be suited to conifers, while low fertility in lowland regions may prevent the cultivation of high yielding coppice species such as poplar and willow.

Where land is released for multipurpose crops the species choice is restricted to those crops which provide the required outputs. Similarly other simultaneous uses of the land such as landscape, amenity, game rearing, recreation and water catchment may place restrictions on crop choice.

The duration of land availability will determine whether a catch, annual or perennial crop is grown. The nature of the opportunity, duration of land availability and environmental factors provide one possible basis for the classification of fuel crops.

Fuel quality

An alternative view would be to classify fuel crops on the basis of their outputs, although, in some cases, a number of different products can be obtained from the same feedstock via different conversion processes.

Biomass is capable of providing solid, liquid and gaseous fuels, but it seems likely that, in the future, a premium will be placed on liquid fuels. The major liquid fuels provided by biomass are vegetable oils, ethanol, methanol and the hydrocarbon mixtures resulting from direct liquefaction and pyrolysis.

Oilseed rape would appear to be the most appropriate fuel crop for vegetable oil extraction which can be cultivated in the UK, but Spedding *et al* (1979b) suggest that the castor oil plant, sunflower and poppy may also have potential as oilseed crops. Ethanol could be produced from sugar beet or starchy crops such as potatoes and cereals or from the Jerusalem artichoke. Methanol is likely to be most efficiently produced from the thermal processing of wood. Catalytic liquefaction may be able to operate on any type of biomass.

Gases can be derived from wet, green feedstocks via anaerobic digestion or pyrolysis or gasification of wood and other dry feedstocks. Wood, straw etc also provide solid fuels suitable for burning.

In general, trees provide the most versatile feedstock capable of high yields in the British climate and of providing a range of solid, liquid and gaseous fuels. Their major disadvantage is that they require long-term dedication of the land to fuel production.

Genetic resources

Apart from oil extraction and ethanol production, where a high yield of a particular chemical fraction of the plant is desirable, most biofuel production systems would aim at maximum (cellulosic) biomass production. Given any specific requirements of the conversion process and the constraints related to the land opportunity, species will be selected on the basis of high yield in the particular environment of cultivation.

Initially fuel crops will be drawn from amongst conventional agricultural and forestry species, but as the criteria for fuel crop selection differ from those of food or forestry crops there is scope for breeding and selection towards the specific requirements of the fuel crop system and for innovation.

Indigenous and introduced species growing wild in the UK and exotic crops, may provide alternative fuel crops. The former have been investigated at the Institute of Terrestrial Ecology, Merlewood Research Station, and a number of indigenous and naturalised wild species, apparently capable of very high yields, have been isolated (Callaghan *et al*, 1981).

CATCH CROPS

The most promising agricultural species on the basis of yield for cultivation as catch crops are Fodder radish *(Raphanus sativus)*, Stubble turnip *(Brassica rapa)*, White mustard *(Sinapis alba)* and rape *(Brassica napus)*. There is a range of varieties available within some of these species, notably stubble turnip. The indigenous weeds Common hemp nettle *(Galeopsis tetrahit)* and Sterile brome *(Bromus sterilis)* and the South American crop Quinoa *(Chenopodium quinoa)* may also have potential as catch crops (Spedding & Carruthers, 1980).

DEDICATED FUEL CROPS

Herbaceous species

Spedding *et al* (1979b, 1980) surveyed a large number of agricultural species and concluded, on the basis of productivity, suitability to the UK environment, costs and suitability for conversion to a fuel, that ryegrass, Red clover, lucerne and Forage rye were the most suitable break or plantation fuel crops and, additionally, sugar beet was suitable as a plantation fuel crop. They envisaged that 'break' fuel crops would be grown in preference to the relatively unprofitable breaks often necessary on cereal farms, while 'plantation' fuel crops might replace many kinds of food and feed crops, possibly being grown on a larger scale and separate from traditional farms. Both 'break' and 'plantation' crops were seen as long-term opportunities.

Of the indigenous and naturalised species considered by Callaghan *et al* (1981) annuals such as Policeman's helmet *(Impatiens glandulifera)* offer an alternative to agricultural annual fuel crops. Perennials such as Japanese knotweed *(Reynoutria japonica)*, apparently capable of annual production in excess of 20 t per ha in small stands, and Giant hogweed *(Heracleum mantegazzianum)* provide an alternative to agricultural perennials or trees. Such perennial crops would require a similar long-term commitment of the land to that required for trees.

Trees

There is an appreciable number of tree species with potential as fuel crops. The choice of systems will depend on the system of management (coppice or single-stem), the climate and the soil fertility. Promising species for short rotation coppice, on the basis of high biomass yield, include species of poplar, willow, alder, *Nothofagus* and of the exotic genus *Eucalyptus* (Pearce, 1983). Species being considered for single-stem forest-energy plantations include Black alder, Pubescent birch, *Nothofagus procera*, sycamore, Hybrid larch, Scots pine, Corsican pine, Douglas fir, Western hemlock and Sitka spruce (Mitchell, 1982).

5 Implementation

Chapter 3 outlined a number of alternative strategies which might be used to increase the amount of fuel produced from biological materials in the UK. These strategies differ in their potential, not only in terms of the types and quantities of fuels produced, but also in the likelihood of their being implemented. It is now important to consider:

(i) the constraints which might act on the implementation of these strategies;

(ii) the implications of large scale development of biofuel production at the national and regional level, and the impact on agriculture and employment; and

(iii) the potential for government action to encourage biofuel production.

CONSTRAINTS ON BIOFUEL PRODUCTION

Economic constraints

The primary constraint on the development of biofuel production in the UK is likely to be economic. If it is to be widely introduced, fuel production must be more profitable than alternative uses of the land, feedstock or fuel. Several feedstocks have alternative uses which may be more profitable than conversion to fuel in particular circumstances. For example, straw may be used in paper and board production, for livestock bedding, as animal feed, for thatching, or to produce furfural (a chemical feedstock) (Staniforth, 1982). The fuels themselves may also have higher value in other markets. Sheppard & Lipinsky (1983) suggest that production of pharmaceuticals, food, feed, and basic chemicals, may be more attractive than use as a fuel. The size of these alternative markets, however, may be rather small and localised; thus in many areas fuel production is likely to be the main use of feedstocks and fuel.

Within agriculture, fuel production will be in competition with existing enterprises (some potential fuel production systems may not displace existing enterprises, but will still be competing with other farm activities for investment). Some types of fuel crops may, therefore, be introduced even if their profitability is quite low. This may be due to the lower profitability of alternative crops, or because they are not in direct competition with existing crops for resources. The production of wood energy is attractive now in large areas of upland Britain, because the returns on existing extensive livestock production are so low (Brandon *et al,* 1983). Catch fuel crops, unlike dedicated fuel crops, are generally not in competition with existing crops, and thus need not be as profitable as them, though they must cover the opportunity cost of the change from winter to spring sowing of the subsequent crop. Even a nominally unprofitable process may be adopted if it saves other costs, for example the value of anaerobic digestion as an anti-pollution measure (ADAS/BABA, 1982).

On most farms at present, though, fuel production is not able to bid resources away from existing enterprises (sample costings for several biofuel production systems are presented in Appendix 2). Improvements in the relative profitability of most biofuel production systems will therefore be necessary if they are to be widely introduced on farms. However, a major transfer of land to fuel production, or use of food crops as fuel feedstocks, may reduce national food production. Unless there are food surpluses or the shortfall can be made up by equivalently priced imports, food prices are likely to rise. Thus the widespread development of fuel production may actually reduce its competitive position. The relative profitability of biofuel production is thus likely to be in dynamic equilibrium with the alternative uses of the land, feedstock and fuel.

It is therefore apparent that the profitability of biofuel production is likely to change over time (both upwards and downwards) and will vary widely between different fuel production systems. The profitability is a function of three factors: feedstock production costs; feedstock conversion costs; and the value of the fuel produced. Each of these is now considered in more detail.

Production costs

Agricultural residues are likely to produce the cheapest feedstocks. Their production costs are largely borne by the main enterprise, and there may also be costs associated with their disposal which may be avoided if they are used as a fuel. In addition, fuel feedstocks from multipurpose crops will not have to bear their full production costs, as these will be divided between the different components. For fuel production systems, however, where the crop is specifically grown for fuel, the full production costs must be borne by the fuel feedstock and may be a significant proportion of total costs (Hall, 1982).

Several species proposed as fuel crops are already grown in existing systems. It would seem likely that production techniques for fuel cropping systems will be broadly similar to existing practices, thus their production costs may be expected to be comparable. New species specifically grown for fuel production should have lower production costs than using existing species as fuel crops. The introduction of more efficient production systems, referred to in Chapter 3, may also be expected to reduce feedstock costs.

The total production cost for the feedstock will include all costs up to the point of conversion to fuel. It is thus likely to include additional elements such as storage and transport charges. Some particular fuel production systems may also have specific additional costs which must be included in the total. Enclosure of areas in the uplands for wood energy plantations, for example, is likely to incur high fencing costs, which may make small-scale afforestation uneconomic.

Biomass production is generally highly seasonal and may not match demand, thus some storage of feedstocks is likely to be necessary. Wet biomass feedstocks are likely to undergo rapid deterioration. Anaerobic digestion, the main conversion technique for such feedstocks, is a fairly slow process, thus unless operated with excess capacity for most of the year, it is unlikely to be able to cope with seasonal peak production. Some form of conservation, such as ensiling, will therefore be necessary to store feedstocks prior to conversion. Conservation is likely to be costly, and may also lead to a reduction in the quality of the feedstock for fuel production. As heat and electricity cannot be efficiently stored, feedstocks for their production must be stored until needed. Dry biomass feedstocks for direct combustion may be fairly easily stored, but given the low density of biomass, provision of weatherproof storage to keep feedstocks dry may be a significant cost. Systems such as grain-drying with straw, where demand and supply are closely linked, may therefore be more attractive than systems where storage is required.

Their low energy density also means that the minimum transport of biofuel feedstocks is desirable. On-farm processing is therefore likely to be preferred to off-farm conversion. Some conversion processes, however, may be most economically operated at a scale which may exceed the production capacity of most farms. Central processing of fuel feedstocks may therefore be necessary, if feedstock production potential is to be realised. The feedstock transport costs for a central processing plant will increase with the size of the catchment area, and will indicate an optimum size for the plant (Dunn and Haskew, 1979). A high feedstock production intensity in the vicinity of a processing plant will therefore be desirable to minimise the catchment area. Feedstock densification by compaction or pelletisation may also be used to reduce transport costs. If fuel crops are to be produced in the more remote upland areas of the UK it may also be necessary to improve the road network in these areas to enable the feedstocks to be collected. The successful introduction of biofuel production systems will therefore depend upon the identification of opportunities where feedstock production potential can be matched to economically scaled conversion plants.

Conversion costs

The costs of feedstock conversion would seem the most likely area for improvements in biofuel profitability. Many technologies are still at the prototype or even experimental stage. Their costs when commercially produced and operating at optimum scale may be expected to be considerably lower. Some biofuels are produced at present using adaptations of existing technologies such as beverage alcohol production, which, though highly developed, may be directed toward optimising qualities such as purity, which may be inappropriate for fuel production. Current research is likely to lead to new technologies which are cheaper or have higher conversion efficiencies. It would therefore be realistic to assume some improvement in feedstock conversion costs in the future as a result of technical development, though it would be unwise to rely on new technological advances to justify otherwise uneconomic processes.

Scale of conversion process may affect conversion costs. Several conversion technologies, such as fermentation and distillation and gasification may be most efficiently operated on an industrial scale. Industrial biofuel production plants, however, will require feedstock collection and transport over a fairly large catchment area, which will increase feedstock costs, whereas small-scale production on the farm may be less efficient.

For some conversion processes additional feedstock pre-treatment costs may be necessary. Dry feedstocks such as straw, bracken and reeds, may need to be densified into pellets or briquettes prior to burning, and wood may require chipping to facilitate handling and efficient combustion. Automatic stoking of furnaces may also be considered desirable to reduce labour requirements. Maceration of crop feedstocks for continuous anaerobic digestion is likely to be essential (S P Etheridge, personal communication).

Conversion costs may also have to include a charge for storage of fuel. Liquid biofuels may be stored in conventional fuel tanks, though the lower energy density of the alcohols will require increased storage volume. Methanol, and to a lesser extent ethanol, may require specialised storage to overcome the problem of corrosion and the dissolving of plastic and rubber components. Biogas may be stored in pressurised cylinders, an energetically and economically expensive process, or in unpressurised gasometers. Any gas storage must meet stringent safety standards, which may be costly to implement.

Valuable byproducts may be produced by a conversion process which may be used to offset the costs of fuel production. However, as the study of Webb (1981) indicates, the impact of a major programme of biofuel production on the market for a biofuel byproduct such as distiller's dried grains would be quite significant. Thus the credit attributable to the byproduct use is likely to be reduced if biofuel production is developed on a large scale.

Value of fuel produced

The price a biofuel can command is determined by the price of alternative fuels. At present costs, this does not make biofuel production sufficiently profitable to be introduced on farms. Real increases in fuel prices would thus make biofuel production more competitive. Prediction of future price levels for fuels is therefore a critical assumption in determining the potential development of biofuel production. Pearce (1981) suggests, from a review of recent literature, that a 3% per annum real rise in oil prices to the year 2000, is a reasonable estimate. The UK Department of Energy uses a similar rate of increase in its 'exploratory scenario' (Department of Energy, 1979). It is thought that the rise will not be the same for all fuels. Coal and gas prices are expected to rise more slowly, thus their substitutes will have to be cheaper than equivalent liquid fuels. A detailed analysis of the economics of different fuel production systems under different scenarios of future energy prices is beyond the scope of this paper. Table 5.1, however, indicates the relative costs of a number of biofuels in comparison to conventional fuels at present day prices.

The effects of real increases in national fuel prices will not be limited to the value of biofuels, however. They will also increase the costs of biofuel production inputs and thus reduce the net effect on profitability, and induce other changes throughout the economy which will alter the circumstances for biofuel production. Increasing fuel prices may also be expected to encourage energy conservation and the development of more energy efficient systems which will lead to a reduction in demand (Pearce, 1980) and thus slow the rate of fuel price increases. Other alternative energy sources may require smaller increases in prices to become profitable and may be adopted before biofuel production, thus increasing energy supplies and further slowing price increases. Fuel prices are also influenced by exogenous factors such as the level of prices set by the oil producing nations and the rate of taxation applied by government on fuels.

The value of fuels will vary depending on the type of fuel, and the particular demand for that fuel. Value may be put on particular attributes of a fuel such as cleanliness, or convenience; or on considerations such as security of supply. Liquid fuels for transport are likely to be at a premium, especially in areas dependent on imported oil. Several liquid biofuels are available, but they tend to be expensive to produce. Ethanol and methanol (up to 3%) are attractive as fuel additives as they can be used in conventional engines and can substitute for ethyl lead as octane boosters. Pure ethanol and methanol, however, require some changes to the engine and fuel supply system, and may be difficult to cold start. Pure vegetable oils may be used as diesel substitutes, but produce heavy carbon deposits in the engine. This may be alleviated to some extent by chemical treatment (such as transesterification) of the oils, or by chemical additives, but there are still problems of lubricating oil dilution and sludge formation (Bandel & Heinrich, 1983). An alternative approach would be to modify the engine, or to use a different type of engine, such as the Stirling engine (Morgan, 1969), so that vegetable oils may be used without modification.

Gaseous fuels are difficult to store, but can be used to provide heat, or, in a gas engine, motive power or electricity as required. Electricity may also be produced *via* direct combustion. It is a clean, versatile and convenient fuel, but is inherently inefficient to produce, and must be carefully matched to demand, as it cannot be conveniently stored. Space heating by direct combustion, or burning of gaseous fuels is the simplest use of biofuels, and is the only product of aerobic decomposition. Applications for space heating on farms, however, may be limited. Thus unless new activities to make use of the heat are introduced, or a local market for low grade heat can be found, its value is likely to be low. The value of fuels may therefore be expected to depend to a large extent on their utility.

Though a general ranking of fuel values may be made (ie liquid fuels and electricity are more valuable than gaseous fuels which are more valuable than sources of low grade heat), the actual value of fuels will vary depending on the specific local demand for the fuel, and the availability of alternative sources of supply. Thus if there are no opportunities for the use of low grade heat on-site (as it cannot be easily transported), for example for domestic or glasshouse heating, then heat production systems will be valueless. There are also likely to be variations in the local availability of other fuels, for example high cost, or even absence of supply, of fuels (particularly electricity) in remote rural areas and islands. Systems capable of local production of fuels may thus be particularly valuable. It is therefore difficult to make a generalised assessment of the viability of a biofuel production system, as this may be dependent on the circumstances at any particular site.

Social and organisational constraints

Some of the biofuel production strategies proposed will require social changes. The development of biofuel production may, therefore, be constrained by resistance to these changes. Nationally, dietary change may be expected to be a

Table 5.1

Cost of biofuels relative to conventional fuels

Fuel	Feedstock	Country	Date	Conventional fuel	Relative price of biofuel	Source
Biogas	crop	New Zealand	1979	petrol	1.29–1.48	Harris *et al* (1979)
Biogas	bracken	UK	1981	propane	2.00	Appendix 2 of this paper
Biogas	catch crop	UK	1981	propane	0.92	Appendix 2 of this paper
Biogas	pig waste	UK	1981	propane	0.75	Appendix 2 of this paper
Biogas	green crops	West Germany	1982	oil	1.5–3.8	Kleinhanss & Meister (1983)
Electricity	biogas	West Germany	1980	electricity	1.7	Dohne (1981)
Electricity	biogas	US	1982	electricity	0.33–1.12	Scheller (1983a)
Ethanol	wheat	Australia	1979	petrol	2.03–3.33	Buckland & Buick (1980)
Ethanol	sugar beet	Australia	1979	petrol	1.41–2.70	Buckland & Buick (1980)
Ethanol	wheat	Australia	1979	petrol	2.20	Roberts *et al* (1980)
Ethanol	beet	New Zealand	1979	petrol	1.48–2.24	Harris *et al* (1979)
Ethanol	maize	New Zealand	1979	petrol	1.71–2.52	Harris *et al* (1979)
Ethanol	wood	New Zealand	1979	petrol	1.71–2.29	Harris *et al* (1979)
Ethanol	na	Sweden	1980	petrol	2.5	Bergman (1981)
Ethanol	maize	US	1980	petrol	0.84–1.12	Hertzmark *et al* (1980)
Ethanol	maize	US	1980	petrol	2–5	Lockeretz (1982)
Ethanol	beet	Australia	1981	petrol	1.63–1.97	Wonder & Simpson (1982)
Ethanol	wheat	Australia	1981	petrol	1.97–2.07	Wonder & Simpson (1982)
Ethanol	na	West Germany	1982	petrol	1.79	Meinhold & Kögl (1983)
Methanol	wood	New Zealand	1979	petrol	0.95–1.57	Harris *et al* (1979)
Methanol	bracken	UK	1981	fuel oil	1.63	Appendix 2 of this paper
Vegetable oil	oilseed rape	West Germany	1979	diesel	1.8–2.4	Buhner & Kögl (1981)
Vegetable oil	na	Sweden	1980	diesel	1.7	Bergman (1981)
Vegetable oil	oilseed rape	Australia	1981	diesel	1.21–2.73	Wonder & Simpson (1982)
Vegetable oil	sunflower	Australia	1981	diesel	1.80–2.58	Wonder & Simpson (1982)
Vegetable oil	soya bean	West Germany	1982	diesel	1.8	Bandel & Heinrich (1983)

Table 5.1 (Continued)

Fuel	Feedstock	Country	Date	Conventional fuel	Relative price of biofuel	Source
Bracken	N/A	UK	1981	fuel oil	0.44	Appendix 2 of this paper
Crop residues	N/A	US	1980	coal	1.11–1.36	English *et al* (1981)
Straw	N/A	Sweden	1980	oil	0.53–1.00	Bergman (1981)
Straw	N/A	UK	1981	fuel oil	0.31	Appendix 2 of this paper
Straw	N/A	France	1982	oil	0.21–0.51	Pellecuer (1983)
Straw	N/A	UK	1982	oil	0.51	Staniforth (1982)
Willow	N/A	UK	1982	oil	0.13	McLain (1983b)
Willow	N/A	UK	1932	coal	0.21	McLain (1983b)
Wood	N/A	France	1982	oil	0.16–0.32	Pellecuer (1983)

slow process. At the farm level farmers may be reluctant to change to new crops or production systems. In particular, fuel cropping systems with a longer than annual production cycle may not be readily adopted, as farmers do not normally consider woodland as being within agricultural management systems, and some have a distinct aversion to forestry (DART, 1983). Such 'forestry' fuel production systems would, therefore, seem likely to be managed separately from agricultural systems, though there may be opportunities for small-scale coppicing, for example for domestic fuel, on appropriate sites on farms.

Strong social pressures are likely to act against fuel production systems which affect amenity use of the countryside. If biofuel production led to major changes in landscape, for example, due to large-scale afforestation or the introduction of exotic plant species, this would be likely to be resisted. More specifically, restriction of recreational access to the countryside or interference with fishing (both major national leisure interests) is likely to be strenuously opposed. Biofuel production may also have indirect effects on amenity, for example through the increase in traffic on rural roads.

In general, however, the adoption of fuel production at the farm level may be expected to depend on the ease with which it may be integrated with existing practices and the size of the economic incentive. The use of an existing crop as a fuel feedstock with off-farm processing would not be essentially any different from present practice. Agricultural residues are often already handled in conventional systems, and their diversion toward fuel production should be fairly easily integrated into existing practice. They are likely to be processed on the farm, as they have very low economic value, thus their use in fuel production, if profitable, is likely to be constrained by the availability of labour for their collection and use. Catch cropping is another established practice which may be adapted for fuel production. Its introduction is also likely to be constrained by the availability of labour and machinery, as well as a sufficient growing period within the crop rotation. New crop species for fuel production, if they require little change to existing practices, may also be fairly readily introduced, as the rapid expansion of oilseed rape area in the 1970s indicates.

Some proposed fuel production systems, however, will require significant modifications to farming practice. On-farm fuel conversion is likely to have new labour and skills requirements. New crops and production systems may also require new skills and machinery. These systems are only likely to be introduced if they are significantly more profitable than existing systems, though the rapid adoption of silage making in the past fifteen years suggests that such changes are not impossible.

On-farm fuel production may also be constrained by the range of potential uses of biofuels on the farm. Matching of fuel supply and demand is likely to be a critical influence on the economics of biofuel use. Additional enterprises to make more efficient use of biofuels may therefore be considered, and these also would have to be integrated into the farming system.

There may also be organisational constraints at a national level. If fuels are to be produced at an off-farm processing centre, it may be necessary to operate an area quota system to ensure feedstock supplies, or for the processing plants to be run by producer co-operatives. A large-scale biofuels programme will also require a national distribution and marketing system. The development of such structures will be necessary if fuel production is to be rapidly adopted.

Legal and institutional constraints

The scope for the adoption of fuel production on many farms may be limited by a variety of legal and institutional constraints. These mainly apply to changes in the use of land. This is particularly likely to apply to the development of 'forestry' fuel crops on presently unafforested land. The constraints on land transfer may be as a result of direct ownership of the land, by designation, or by legal rights of non-owners. Five types of constraint may be distinguished:

(i) Legal constraints: commoner's and, in Scotland, crofter's rights are likely to prevent widespread cultivation or afforestation of common or crofting lands. In addition, only 68% of agricultural holdings in GB are owner-occupied (MAFF, 1983). On tenanted farms there may be restrictions in the tenancy on afforestation or use of farm woodland for fuel. A very small area of land may also be affected by restrictions under Capital Transfer Tax exemption agreements. Another type of legal constraint may apply to the introduction of exotic species as potential fuel crops. Under the Wildlife and Countryside Act, 1982 the cultivation of Giant hogweed *(Heracleum mantegazzianum)* and Japanese knotweed *(Reynoutria japonica)* (two of the species considered by Callaghan *et al* (1981), as potential fuel crops) is illegal.

(ii) Wildlife conservation constraints. land in nature reserves, or designated as a Site of Special Scientific Interest is likely to be unavailable for fuel production.

(iii) Amenity and conservation constraints: these cover a wide variety of land uses. The main areas are the designated National Parks, Heritage Coasts and Areas of Outstanding Natural Beauty in England and Wales, and the National Scenic Areas and proto-National Parks in Scotland. Other areas likely to be affected include Long Distance Footpaths, Country Parks and Woodlands, Forest Parks. Stately Homes and Gardens, and Woodland and National Trust Lands. Ancient Monuments and important archaelogical sites may also restrict land use change.

(iv) Water catchment constraints: Water Authorities have expressed concern at the impact of land use change in water catchment areas.

(v) Ministry of Defence lands: the Ministry of Defence would be likely to restrict the transfer of lands which they own and over which they have rights.

The severity of these constraints will be variable: some will totally preclude fuel production, whereas others may be overcome with sensitive planning and consultation. It is therefore apparent that even if fuel production could be both profitable and feasible in some areas, it may not be adopted.

Technical constraints

The main technical constraints apply to the availability and reliability of the technologies for feedstock production and conversion. The low energy density of many biomass feedstocks makes on-site conversion desirable. Thus simple, cheap and robust farm-scale production equipment, as well as large-scale, high efficiency industrial plants, will be needed if fuel production is to be widely adopted.

The reliability of technologies is likely to be an important consideration. If new technologies are unreliable a reserve of conventional fuel is likely to be necessary, probably at high marginal cost, to ensure supplies. Farmers are also likely to place a high premium on reliability of machinery and equipment. A failure at critical times of the year could lead to high costs for other enterprises which are disrupted, as well as the direct costs of repair, and loss of income from the fuel production system itself. For example, delay in fuel crop harvesting could lead to loss of yield in the subsequent crop, or labour might have to be withdrawn from other activities to repair fuel production equipment. For industrial-scale fuel production plants, a reliable supply of feedstock, to ensure optimum capacity operation, and reliable conversion technologies, are likely to be essential.

Though many fuel feedstocks can be satisfactorily, if perhaps not very efficiently, harvested using existing machinery, others are likely to require new technologies. There are a number of prototype machines which have been developed to improve the efficiency of biofuel harvesting and to handle new biofuel crops. For example, the whole-crop harvesting system described by Lucas (1978) allows more efficient collection of straw, which is then available for on-site grain drying. Prototype machines have also been developed for harvesting trees for energy (Keville, 1982; McLain, 1983a). These machines, however, are not widely available and are likely to require considerable development before they are commercially viable.

Similar problems exist with feedstock conversion equipment. Table 5.2 illustrates the state of development of the various conversion technologies considered in Chapter 2. Furnaces for direct combustion of solid biomass fuels are commercially available at industrial- and farm-scale today, and new and more efficient designs are being developed. Farm-scale anaerobic, digestion in the UK is not yet a fully proven technology. All the 'commercial' units are prototype designs, and a number of designs are still at the experimental stage. Industrial applications of anaerobic digestion are still limited to sewage works. No other technologies are yet available for farm-scale or industrial fuel production from biomass in the UK. In other countries such as France and the US, however, biomass conversion technologies are more developed, and commercial anaerobic digestion, fermentation and gasification plants are in operation.

Table 5.2

Status of biomass conversion technologies

Product	Present[1]	Short-term[2]	Long-term[3]
Heat	Direct combustion	Aerobic decomposition	
Gaseous fuels	(Anaerobic digestion)	Anaerobic digestion	
	(Gasification)	Gasification	
		Pyrolysis	
Liquid fuels	(Fermentation)	Fermentation	Cellulosic fermentation
		(Hydrolysis + fermentation)	Hydrolysis + fermentation
		Gasification to methanol	Liquefaction
		Pyrolysis	

1 Commercial at present in UK (viable systems in operation outside UK).

2 Technologies at pilot plant stage expected to become commercial in the short-term (20 years).

3 Experimental technologies operating at laboratory scale — long-term commercialisation (20 years +).

The early stage of development of most fuel production technologies in the UK is therefore likely to discourage farmers from adopting fuel production at present. If fuel production is to be widely introduced on farms, then farmers have to be able to purchase standardised and reliable fuel production equipment as easily as other agricultural machinery. This will require considerable further improvement of fuel production technologies and the development of a skilled and well equipped sales, and repairs and maintenance service.

The large scale development of biofuel production, aimed at supplying national fuel requirements will also need an infrastructure for the distribution and use of the biofuel. Some biofuels such as ethanol if used as a petrol additive may be readily integrated into existing distribution systems, and require little modification of equipment for efficient use. However, others such as biogas will probably require new distribution systems, and adaptation of equipment for its use. These systems will have to be developed before such fuels can be widely used.

Energetic constraints

It has been suggested that another factor in evaluating potential fuel production systems may be the energetic balance. In agricultural systems energy is used to produce food and other commodities, the output of which is normally evaluated on qualities other than energy content. Thus energetic considerations may be of limited value in comparing alternative agricultural systems. In biofuel production, however, the output is fuel and the energy balance would appear to be an important factor.

All the estimates provided in Chapter 3 for the potential output of different fuel production opportunities are gross figures and do not take account of the energy required to produce the fuel. This is true also for fossil fuels where the energy values commonly quoted are the gross energy content of the fuel, and do not include the energy cost of extracting, refining and transporting the fuel. Energy Analysis (EA – see Appendix 3) may be used to derive the Net Energy (NE) produced by a fuel production process. A system with a high NE would seem preferable to a low NE system, and a negative NE would be useless.

There are, however, a number of drawbacks to the use of EA in evaluating potential biofuel production systems, which are inherent in the technique. The aggregation of all energy inputs takes no account of the differences in quality of different primary energy sources, and thus may give an inaccurate impression of the viability of a system. One of the main advantages of biofuels is that they include liquid fuels which may be substituted for petroleum derivatives as transport fuels. A biofuel production system using low-grade biomass feedstocks to provide the process energy for the production of a high-grade liquid fuel therefore, may be considered worthwhile even with a low or perhaps negative NE. What is needed is not energy, but substitutes for high energy density fuels at present derived from non-renewable sources; most biofuel production systems have a positive energy balance in respect of fossil fuel use and are therefore attractive as fuel sources.

Energy analysis also does not have a mechanism to account for intertemporal preference, thus it is unsatisfactory for use in investment appraisal. The problems that exist over the convention to be used to account for labour inputs in EA, also mean that systems with significantly different labour requirements cannot properly be compared. Energy budgets are thus heavily dependent on the assumptions and conventions used in their calculation.

The assumption in EA that energy balance is the critical factor in system selection is also unlikely to be true in reality. Economic criteria are likely to override energetic considerations. As indicated before, the economic value of feedstocks and fuels will be subject to significant geographical and temporal variation depending on the particular circumstances of feedstock supplies and fuel demand and will be affected by exogenous factors such as government policy. Thus the same biofuel production systems may be evaluated differently according to the local conditions, and energetically less efficient systems may be preferred because the fuel produced is more valuable. Fuel may also be produced as a byproduct of some other process which may require sub-optimal energy efficiencies, or may confer non-energy benefits, for example pollution reduction, which may make a poor energy return worthwhile. In choosing between alternative production systems for the same fuel, energy efficiency may also not be the most important criterion – thus improvements in NER may be associated with increasing system complexity, or may require an infeasible increase in scale of production.

However energy analysis can provide useful information in assessing potential fuel production systems. The low energy density of biofuel feedstocks, and the consequent high fossil fuel energy cost of transport, serve to underline the advantage of on-site feedstock conversion and may give an indication of a suitable limit on the catchment area of central processing plants. EA may also help to identify other areas of particularly high energy requirement which may be of particular value when seeking to improve energetic efficiency, or to consider the likely effect of increased fuel prices on the system. NE values for the output of different biofuel production strategies may therefore be useful as an additional measure in comparing their potential, but is unlikely to be the sole or indeed primary factor in such an assessment.

There are, therefore, a number of factors which tend to slow or to restrict the adoption of fuel production systems. Most of these factors are common to any new enterprise and thus do not represent a particular constraint on fuel production. If fuel prices continue to increase, as most studies suggest, then biofuels are likely to become increasingly attractive, and if there is a sufficient financial incentive, then many of these constraints are likely to be eased. Thus whilst biofuel production is unlikely to fulfil all the potential opportunities described in

Chapter 3, the constraints outlined above do not represent an insurmountable barrier to the development of biofuel production from UK agriculture.

IMPLICATIONS OF BIOFUEL PRODUCTION

One of the main advantages of biofuel production is the potential for 'home production' of fuels. This will apply at the national, regional and farm level. The security and convenience of on-site fuel supply may be a positive consideration in evaluating the potential of fuel production. At a national level the stability of fuel supplies may be increased by a reduction in imported fuels. The site specific nature of fuel production viability may make it locally important in some regions. Locally produced fuels may be particularly valuable in isolated rural or island communities at the periphery of existing distribution networks. At the farm level, farmers may also be attracted by the opportunity of developing a degree of self-sufficiency of fuel supplies (Smith, 1980).

National policy implications

Unless there are substantial improvements in food and timber production efficiency, or fuels are to be produced solely from wastes, fuel production is likely to be in competition with existing forestry and agriculture for land resources. The large-scale development of biofuel production in the UK is therefore likely to have a significant impact on UK agriculture and forestry. Some coordination of national policy on the alternative biomass-using systems would therefore appear desirable. Coordination of food, fuel and timber production would seem likely to require an appraisal of land use policy.

The increased pressure on land resources from a major biofuels programme would make some control on land transfers, particularly to urban or other biologically non-productive uses, desirable. Measures to avoid the inefficient use of land, for example in the urban fringe, and to encourage the reclamation of derelict and despoiled land would help to increase the total productive biomass available. A coordinated policy on biomass use may also be expected to encourage the development of integrated fuel/food and fuel/timber production systems, similar to those developed on some upland farms (Mutch & Hutchinson, 1980) which may be more profitable than the separate systems. Fuel production might also be integrated with recreational use of the land. The widespread adoption of biofuel production could therefore lead to a greater efficiency of land use through: intensification on existing lands; use of presently unused or under-used land; and integration of activities.

At present a major aim of UK agricultural policy, as stated in the 1975 White Paper 'Food from our own resources' (MAFF, 1975) and reiterated in 'Farming and the Nation' (MAFF, 1979) is an increase in national food self-sufficiency. An increase in indigenous timber production is also proposed (Forestry Commission, 1978; Fairgrieve, 1979; CAS, 1980a). However, increasing national self-sufficiency may not be the most efficient way to ensure national supplies (Ritson, 1980). If biomass resources are to be efficiently allocated then an assessment of the relative strategic values of food, fuel and timber production to determine the optimum national level for each would appear desirable. Such an assessment would involve considerations, not only of economic efficiency, but also factors such as stability and security of national supplies and commitments to the EC and international trading agreements. The situation within the EC is obviously different from that in the UK alone, and priorities are likely to be different. A biomass policy for the UK, therefore, ought to be considered within the wider context of the EC. At present, however, UK commitment to the EC is somewhat equivocal, thus purely national considerations are also likely to influence policy choice.

Implications for agriculture

The development of large scale production of fuel from biomass in the UK is likely to have implications for all sectors of agriculture. The greatest impact is likely to be in those parts of the industry, such as horticulture, where fuel forms a high proportion of total costs, and where feedstocks for fuel production are cheaply available. For example, the anaerobic digestion of wet crop residues, such as tomato haulm, or the direct combustion of dry wastes could help to provide space heat for glasshouse heating. This might also encourage the development of fuel using activities in areas where feedstocks are cheaply available, for example glasshouse production in East Anglia, heated with surplus straw from neighbouring farms (Balls, 1982). For controlled environment intensive livestock units, fuel may be also a substantial cost, and waste treatment to reduce pollution a necessity. Fuel production could thus be a byproduct of pollution control and help reduce costs.

Another area where biofuel production is likely to have a major impact is in the upland regions. Short rotation forestry, coppicing of natural woodland and collection of natural vegetation could provide additional income on marginal farms which could help to revive agriculture in these areas. Large-scale afforestation, such as could be required for wood energy plantations, might lead to the loss or wastage of good agricultural land, and could also deleteriously affect lowland agriculture through a reduction in the supply of store stock (National Farmers Union of Scotland, 1979). Farmers might also prefer to sell up their farms and move to better quality land, rather than afforest their farms. Smaller scale afforestation integrated with agriculture, and encouragement of better use of on-farm woodland, however, could be a positive contribution in these areas.

Fuel cropping is unlikely to be so immediately economically attractive in lowland agriculture. However, it could offer a number of advantages. The development of farm woodland as a source of on-farm or commercial fuel would encourage better management and use of this poorly utilised resource. In urban fringe farming, a fuel crop might be subject to less damage or theft than a food crop. Fuel production where biomass quantity rather than quality is the important factor may also be able to make use of derelict or despoiled land otherwise unsuitable for agriculture, for example due to pollution. Fuel production might also increase the diversity of crops grown, and perhaps encourage new activities such as fish farming using anaerobic digester waste heat. Such diversification could help to increase farm income stability. Fuel production need not necesarily imply a loss of food output from agriculture, but could use land released by intensification of food production. Within the EC at present, there is overproduction of a number of commodities, and there is considerable political pressure to reduce the consequent cost to the Common Agricultural Policy. Fuel crops might provide an alternative on land at present used to produce the surplus commodities, and thus keep the land in productive use.

Employment implications

A major programme of fuel production is likely to have implications for rural employment levels. The impact will vary according to the particular circumstances, but a number of general effects may be discerned. On-farm fuel production would seem unlikely to lead to many more jobs, except perhaps in the uplands, where a revival of agriculture would boost the rural economy. In other areas it may lead to increased incomes and greater efficiency of labour use and thus help to maintain farm employment. Wood energy 'plantations' and the use of forestry wastes for fuel would seem likely to sustain existing forestry employment, and create new jobs directly and indirectly in fuel production. The construction, supply and servicing of fuel production equipment should also offer new employment opportunities. Estimates of the job creation impact of biofuel production programmes have been made in various studies, and these are summarised in Table 5.3.

RECOMMENDATIONS FOR GOVERNMENT ACTION

Biofuels have the potential to make an appreciable contribution to national energy supplies. However, if this potential is to be realised as and when it is needed, then government action is likely to be necessary to encourage the development of biofuel production and to facilitate its implementation. Three main areas for government action may be identified:

(i) research, development and demonstration;

(ii) the extension of government support for agriculture to biofuel production systems;

(iii) changes in government legislation.

Table 5.3

Employment implications of biofuel production and conversion

Country	Number of jobs generated per 1 Mt DM feedstock	
Canada	3100	wood production and conversion to methanol
EC	1900	wood production and pyrolysis
EC	2500	residue and waste conversion
Ireland	3900	short rotation forestry
New Zealand	1350	wood production and conversion to methanol
New Zealand	1450	maize production and conversion to ethanol
Sweden	2500	wood production
US	1360	wood production and conversion to methanol
US	5360	maize production and conversion to ethanol

Source: Adapted from Harris *et al*, (1979); General Technology Systems Limited (1981); Neenan *et al*, (1982); Hall (1983).

Research, development and demonstration

Encouragement, coordination and funding of research, development and demonstration of biofuel systems is a particularly important area of government action if the full potential of biofuels is to be realised. Both a continuation and expansion of current research programmes, and the initiation of new projects are needed. Important areas of research may be identified at all stages of biofuel production and use.

Economic studies

Studies to assess the social and economic implications of biofuel production at a national, regional and local level are of particular importance. The widespread adoption of biofuel production in the UK would add another dimension to the issue of national and EC self-sufficiency in timber and food. A major study is needed to assess the relative strategic value of fuel, food and timber production in order to determine the nationally desirable levels for each. Assessments of the local and regional, social and economic effects of the most promising biofuel production opportunities, such as straw utilisation, on-farm anaerobic digestion, wood energy plantations and catch cropping for fuel are also needed. The markets for particular biofuel conversion technologies, as determined by feedstock availability and energy demand, also need to be explored.

Feedstock production studies

Existing research and development in the following areas needs to be continued and in many cases expanded:

(i) the growth and yield of fuel crops under different forms of management (eg short rotation coppice);

(ii) the growth and yield of new fuel crop species and varieties;

(iii) the productivity of natural vegetation with repeated harvesting;

(iv) the more efficient recovery of wastes and residues (eg whole crop harvesting);

(v) harvesting fuel crops (eg coppice harvesters, machinery for harvesting natural vegetation on steep slopes or in wet conditions).

The requirements of fuel crops may differ from those of food and timber crops, and there is scope for breeding and selection to those specific requirements Biofuel production could be integrated with food and timber production, with other land uses such as recreation, or with land reclamation. Such systems may be environmentally and economically more desirable than monocultures, or single land uses. Research and development of agroforestry and other forms of integrated land use, which offer the opportunity for biofuel production and improvements in overall biomass productivity, would seem important.

There is also a particular need for practical demonstration projects of the most promising forms of fuel crop cultivation and land management. Such projects should include demonstration of: agronomic or silvicultural aspects; engineering aspects related to harvesting and storage; and, if appropriate, on site conversion and fuel utilisation. The estimated potential of wood energy plantations, in particular, merits demonstration of this opportunity.

Feedstock conversion studies

As Chapter 2 indicated, considerable research on feedstock conversion technologies has been conducted. Demonstration of the most promising technologies through government funded pilot plants would now seem important, and has been identified as a priority in a number of reports (Harris *et al*, 1979; Scantland *et al*, 1979). Further research to develop and demonstrate cheap, reliable and simple conversion units and ancillary equipment will be needed if they are to be widely adopted on farms. More fundamental research is also needed on improving some of the novel conversion processes such as direct liquefaction and cellulosic fermentation which are still at a very early stage of development.

Fuel utilisation studies

Although biofuels are intended to substitute for conventional fuels, their characteristics are such that their behaviour is often different. Considerable work, in recent years, has gone into developing efficient straw and wood boilers and furnaces, and a wide range of designs are available. Further research may improve efficiency and ease of management still further. The use of biofuels such as biogas, methanol and vegetable oils is not very successful without fuel and/or engine modification. Thus research is needed to improve the efficiency of utilisation of biofuels in existing engines, and in the longer term to develop multifuel engines and engines specifically suited to their characteristics.

Extension of government support for agriculture to biofuel production systems

In 1981/82 the government spent over £970m on support for UK agriculture (MAFF, 1983). If fuel and food production are to compete on an equal footing then similar support to that applied to agriculture would need to be extended to fuel production. Government subsidies for fuel crop production and use, might be considered appropriate to accelerate the introduction of fuel cropping, or to sustain production in times of crisis. Continued subsidy of uneconomic fuel production systems, however, would not seem desirable, though government support to stabilise the biofuel market may be more acceptable. Indirect government support could also be made available, for example through farmer education, grants for hardware installation or conversion, and tax incentives on biofuel-using equipment. Assistance to overcome the financial problems associated with the introduction of long term 'forestry' fuel production systems into traditionally short term agricultural management systems may help to encourage their adoption.

Changes in legislation

Present taxation policies also discourage the sort of long term investment which is required for 'forestry' fuel cropping systems, thus tax changes to alleviate this problem may be necessary. The present regulations on the distillation of ethanol are also restrictive, some form of modified licence for fuel alcohol production such as the experimenters' licences available in Australia (Wonder & Simpson, 1982) might overcome this problem. Changes in other areas of legislation may also indirectly encourage fuel production. Thus tighter pollution controls would be likely to encourage anaerobic digestion as a treatment for agricultural wastes, and a statutory control of field burning of straw might encourage use of straw as a fuel. Similarly, the use of ethanol and methanol may be encouraged, by measures to phase out lead as a petrol additive.

6 Discussion

Energy price increases in the early 1970s stimulated research on biofuels. This research has been mainly in relation to:

(i) biomass conversion technologies;

(ii) feedstock production and availability:

(a) availability of wastes and unharvested natural vegetation;

(b) availability of land for fuel crops;

(c) productivity and management of fuel crops;

Less attention has been given to:

(i) the overall potential production of biofuels in the UK and a systematic assessment of possible ways in which this potential could be realised;

(ii) assessment of the constraints likely to act on the adoption of biofuel production systems;

(iii) the implications of the implementation of biofuel production systems for:

(a) national policy in relation to energy, fuel and timber supplies and other national objectives;

(b) agriculture;

(c) employment.

This paper has sought to fill that gap.

Biomass is capable of providing a range of solid, liquid or gaseous fuels. Apart from the possibility of generating hydrogen via photo-biological and photo-chemical processes, biomass is the only renewable energy resource capable of providing such storable, chemical fuels; other renewable energy technologies generate heat, mechanical power or electricity. A particularly important feature of biomass is its ability to provide substitutes, such as methanol, ethanol and vegetable oils, for the petroleum-based liquid fuels.

Direct combustion (not strictly a conversion process as it uses the biomass itself as a fuel) and aerobic decomposition produce only heat. The high grade heat from direct combustion could be used for electricity generation. Anaerobic digestion, gasification and some types of pyrolysis produce gaseous fuels. Liquid fuels may be produced directly by fermentation, pyrolysis, direct liquefaction and mechanical extraction, or indirectly via conversion of 'synthesis gas' from gasification.

The biomass conversion technologies in the UK are at varying stages of development and commercialisation. Direct combustion of straw and wood is already economic and commercial systems are available. In the UK, anaerobic digestion still requires further development if it is to be widely adopted. Fermentation to produce ethanol, and extraction of vegetable oils are well established technologies. Their use in fuel production at present, however, is uneconomic, largely due to high UK feedstock costs. The technologies for thermal processing of biomass to produce methanol and substitute natural gas are relatively advanced, but their commercialisation is limited by the availability of suitable markets matched to the large scale of the processing plants which are needed to make it commercially viable. Direct liquefaction and the more novel conversion processes such as cellulosic fermentation still require further research before they are likely to be economic.

Thus there are a number of promising technologies for conversion of biomass to fuel and it is hoped that funds and facilities will continue to be available for research, and to enable the development of cheap, reliable and commercially viable conversion units. Similarly the genetic resources and management systems to grow a range of types of fuel crops are available, and further development and innovation in both these areas may substantially improve productivity.

This paper proposes five possible strategies to make biomass available for use as a feedstock for biofuel production:

(i) utilising wastes, residues and unharvested vegetation. This could provide up to 818 PJ of energy (10% 1981 UK energy consumption) in various forms (mostly solid combustible material or biogas), with only a minor effect on current agricultural practice;

(ii) increasing the total UK primary production by using unused land and by crop yield improvements, thus releasing land for biofuel production. Growing fuel crops on unused land and on land released by improvement in crop productivity could contribute between 842 PJ and 1344 PJ of energy (10-16% 1981 UK primary energy consumption). Implementation of this strategy would involve some changes in agricultural practice and land use;

(iii) reducing the amount of biomass used as animal feed thereby releasing biomass or land for fuel production. The effect of a change in the quality of animal products is difficult to assess, but dietary changes towards plant products could release sufficient land to provide as much as 1738 PJ of fuelwood (21% of 1981 primary energy consumption). Adoption of this strategy would mean more drastic changes both in farming practice and in the national diet;

(iv) decreasing the output of other products and, hence, enabling the transfer of land from current enterprises to biofuel production. This would occur only if this increased profitability and, if widely adopted, would have consequences for the prices of other products and, hence, the competitiveness of biofuels. The transfer of certain areas of grazing land to wood energy plantations appears profitable now and capable of providing between 320 PJ and 760 PJ of fuelwood (4-9% 1981 UK primary energy consumption);

(v) increasing reliance on food imports and so releasing land for fuel crops. This may occur in response to a bioclimatic or economic rationalisation of agricultural production within the UK or EC.

The potential to produce fuels from biomass in the UK is clearly substantial, but the realisation of this potential will be subject to a number of constraints.

The primary limitation is likely to be economic: biofuels will not be introduced unless they are more profitable than existing uses of the land or feedstock. The use of the fuel produced, as a fuel, must also be more profitable than alternative uses, such as as a chemical feedstock. The profitability of biofuel production will depend on the costs of feedstock production and conversion in comparison to the value of the fuel produced. The development of biofuel production may also be constrained by social and organisational factors, by legal and institutional constraints (particularly on land use change), and by the availability and reliability of the technologies for production, conversion, distribution and utilisation of the fuels. The energetic balance of fuel production would seem unlikely to be a significant factor in system choice if the economics are favourable.

Even allowing for these constraints biofuels seem capable of making a worthwhile contribution to national energy supplies. Although widespread adoption of biofuel production is unlikely to occur in the immediate term, as few systems are competitive with existing enterprises, a number of factors seem likely to make biofuel production more attractive in the future. These include:

(i) *Development of biomass conversion technologies.* Further successful development of biomass conversion technologies would make biofuels more attractive and more competitive with other energy systems.

(ii) *Energy price increases.* The price of fossil fuels, in particular liquid transport fuels, is expected to rise; this will increase the value of biofuels.

(iii) *Changes in national energy consumption.* Increased efficiency of energy use and improved energy conservation could lead to an overall reduction in national energy consumption without significant social change and, thus, enhance the proportional contribution of biofuels to national energy supplies.

(iv) *Food surpluses.* Although the conversion of food surpluses to fuel is not a satisfactory long term strategy, the existence of food surpluses indicates that resources are available for, and may be more profitably used in, the production of commodities other than food. Increased pressure against the overproduction of food in the future is likely to make biofuel production more desirable.

In the light of these factors, particularly fuel price increases and the further development of biomass conversion technologies, the future potential contribution of biofuels to energy supplies in the UK (and EC) appears as follows:

(i) in the short term biofuels could make a small contribution to energy supplies *via* the utilisation of wastes and byproducts with only a minor impact on farming practices;

(ii) in the medium term a moderate contribution to energy supplies could be provided *via* wood energy plantations on marginal land, catch crops and the management of natural vegetation. These options would require the acceptance of some changes in agricultural practice and land use;

(iii) in the long term if biofuels are considered a national priority, multipurpose and dedicated fuel crops could make a substantial contribution to national energy supplies. This would require the acceptance of appreciable changes in agricultural practice, land use and diet.

Biofuels also have a number of attributes which may be considered desirable at a national level.

(i) *Indigenous fuel production.* Biofuels increase the potential for indigenous energy production and thus might increase security and stability of national fuel supplies.

(ii) *Diversity and decentralisation of energy supplies.* Biofuels would seem to offer the greatest potential as part of a diverse energy supply system drawing on a number of different energy sources. The localised nature of their production and high transport costs also mean that they are likely to be particularly important in some regions and on specific farms. A national assessment is likely to underestimate their value in these situations.

(iii) *Flexibility of fuel supply.* Biofuels provide a range of different fuels and may be produced in a relatively short time and at a range of scales of production. The level of production of biofuels can thus be fairly readily adapted to national requirements. Biofuel feedstocks can also be used in the production of commodities other than fuel, such as paper and chemicals.

(iv) *Impact on marginal agriculture.* Biofuel production would seem likely to have a particular impact on marginal agricultural land, such as in the uplands, where wood energy plantations are competitive with existing enterprises in many areas at present.

(v) *Impact on employment.* The introduction of fuel cropping may help to sustain, and perhaps in some areas increase, rural employment both directly and through its effect on the rural economy. New jobs may also be created in the supply and maintenance of fuel production equipment.

(vi) *Low energy futures.* Adoption of the objectives behind some of the 'low energy futures' described by the Department of Energy in 'Energy Paper 39' as alternatives to the 'orthodox view' would render biofuel production particularly important, especially in the light of the potential described in this paper. However, biofuels could equally make a very valuable contribution to national energy supplies in more conventional scenarios.

It would therefore seem likely that the introduction of biofuel production in the UK will become increasingly attractive in future. The widespread adoption of biofuel production is likely to have a significant impact on agriculture and forestry. Thus some coordination of national policy on agriculture, forestry and biofuel production would seem desirable. In particular, the introduction of fuel cropping would seem likely to increase the pressure on UK land resources, and thus might encourage some control on land use change. Within UK agriculture the impact of biofuel production is likely to vary between the different sectors, the greatest impact probably being on marginal agriculture in the uplands and urban fringe.

In the short term biofuels are likely to be considered as substitutes for existing fuels and hence might have to be modified to suit existing fuel utilisation and distribution equipment. In the longer term, if biofuel production develops sufficiently, equipment specifically designed to suit the characteristics of biofuels, or capable of using a range of fuels is likely to be needed.

If the potential of biofuel production is to be realised then government action to encourage its development is likely to be needed. There are three main areas where government support to biofuel production would seem desirable.

(i) *Research, development and demonstration.* This would seem to be the most important area for government support at present. Research is needed at all stages of biofuel production and use.

(ii) *Extension of government support for agriculture to fuel production systems.* At present there are a wide range of supports available for agriculture which, if biofuel production is to compete on an equal footing with existing agricultural enterprises, should also be available for biofuel production.

(iii) *Changes in legislation.* Some existing legislation is a constraint on the development of biofuel production. Changes could include tax concessions to encourage investment in 'forestry' fuel production systems, and modified distiller's licences to allow on-farm fuel alcohol production. Legislation in other fields may also encourage fuel production. For example, pollution controls might encourage use of wastes as fuels.

Such measures are unlikely to transform the potential of biofuels overnight, but will help to encourage the smooth development of a viable biofuel production system in the UK. In the past, biofuels were the primary source of energy in the UK. Though it is unlikely that they will regain their former pre-eminence, it is apparent that biofuels could make an appreciable contribution to UK energy supplies in the future. If this potential is to be realised (as and when nationally desirable) then careful planning and management of national biomass resources and a continuing programme of research and development are essential.

7 Recommendations

The stage of development of biofuel production in the UK means that the recommendations of this paper relate to research rather than policy. If biofuels are to make a worthwhile contribution to national energy supplies in the future, then further research is necessary now to ensure that the full potential of biofuels can be realised as and when it is needed.

Considerable research has already been conducted on the technical aspects of biofuel production, and, as this paper has shown, there is an appreciable technical potential for the production of biofuels in the UK. The next stage should be to demonstrate the most promising production systems, and to improve efficiency.

Reliable Information on the economics of biofuel production in the UK is, at present, very limited. In this paper therefore, it has only been possible to outline some of the economic factors which may affect the development of biofuel production. An overall quantitative assessment of economic potential at the national level, such as could be produced from existing data, would be of limited value as the viability of biofuel production systems is likely to be dependent on the particular local circumstances. Economic studies therefore, would be better concentrated on specific opportunities as they are identified.

Some of the implications of biofuel production for UK agriculture and national policies are considered in this paper, but further research in this field and the formulation of a national biofuels policy would seem important.

Five major areas for further research development and demonstration may therefore be identified.

(i) *Economics*

(a) Detailed economic studies of particular biofuel production opportunities, such as wood energy plantations in the uplands.

(b) Exploration of markets for specific technologies (eg on farm or regional anaerobic digesters, gasification plants) as determined by feedstock availability and energy demand.

(ii) *Policy*

(a) An assessment of the relative strategic value of food, fuel and timber production at a national level.

(b) Studies of the social and economic implications of the development of biofuel production at the local, regional and national level.

(iii) *Feedstock production*

(a) Full-scale demonstration of the most promising systems (eg wood energy plantations, bracken management).

(b) Further research on existing and novel fuel crop species to improve yields and ease of management.

(c) Development of harvesting machinery for fuel crops and residues.

(d) Development and demonstration of agroforestry systems and the integration of biofuel production with crop and animal production, and recreation.

(e) Research on integrated systems of production, energy use and recycling of wastes and nutrients.

(iv) *Feedstock conversion*

(a) Pilot plant demonstration of technologies such as thermal processing to produce methanol and substitute natural gas.

(b) Further development of other technologies, such as anaerobic digestion, to produce cheap, reliable and simple systems for use on the farm.

(v) *Fuel utilisation*

(a) In the short term – development and demonstration of the use of biofuels as substitutes for, or additives to conventional fuels, and their use in conventional engines, boilers and furnaces.

(b) In the longer term – development of engines specifically designed for biofuels and of multifuel engines.

Appendix I
Descriptive model of energy flow through the UK biomass system

The flow of solar energy through the UK biomass system is represented in the form of a simple hydrodynamic analogue illustrated in Figure 3.1. The tanks are as follows:

(i) *UK Primary Production* includes all crops, grass and other plants produced within agriculture towards the provision of a biomass product, vegetation managed towards a non-material end (eg recreation) and unmanaged vegetation.

(ii) *Imported Biomass* results from primary or secondary production and is used directly (ie as food, timber or other commodity) or via UK secondary production (ie animal feeds). *Exported Biomass* covers that leaving the country in any form.

(iii) *UK Secondary Production* includes all forms of secondary or higher level production (eg farming of carnivorous animals such as mink, trout etc) occurring in the UK on substrates from UK primary production or imported primary or secondary biomass. The category refers primarily to managed secondary production, but also covers natural secondary production and consumption, the outputs of which are largely 'unused' (except where wild animals are hunted). It would also include industrial food production from imported substrates (eg single cell protein from petroleum), but these latter categories are relatively insignificant.

(iv) *Recycled:* This refers to biomass which, having been consumed by humans in some way is fed back into the system. It covers the recycling of human waste, used timber, paper etc, towards useful products or to animal production.

(v) *Timber, Food, Fuel, Other Commodities* are the useful or consumable products of the biomass system. The categories refer to these commodities at the point of consumption and any material produced but not consumed under any of these headings is automatically 'unused'. 'Other commodities' covers products from primary production such as wool, leather and fur. The demand for some of these products may increase in the future.

(vi) *Unused:* By definition this is the sum of biomass produced which is not consumed as timber, food, fuel or other commodities and post-consumption biomass which is not recycled. It will include, as already stated, the wastes and byproducts of production processes, and biomass grown naturally which is not harvested or used as a consumable product.

(vii) *Consumption* refers to the use of biomass products by humans and includes food consumption, use of timber in construction and furniture, burning of fuels and utilisation of other commodities. The consumer here is the final user and not the industries which process the product prior to consumption.

Use of the model

The numbered valves govern the rate of flow of biomass into the various tanks. Flows will have efficiencies associated with them; as they are the summation of a number of processes with different efficiencies, the overall efficiency depends on the proportions of biomass flowing *via* the various processes. A process will produce primary products and byproducts, which may flow to a different sink from that of primary products, and involve energy losses via non-biomass outputs (eg heat).

At a conceptual level the behaviour of the model is such that adjustment of one valve will lead to a change in the flow through all the valves. However, implementing that adjustment will only be possible in specific terms and will automatically involve adjustment of other valves (eg using arable land to grow fuel crops will increase the flow through valve 7, but also involve a decrease in flow through valves, 5, 3 and possibly 8). The diagram can be used to visualise the overall effect of such changes.

The diagram, therefore, provides a useful basis for planning changes in the allocation of biomass to various products. Increase in the flow of biomass to one product can be compensated for by a planned adjustment elsewhere. Additionally, the intrinsic effects of implementing any specific change can be visualised.

Appendix II
Economics of biofuel production

Estimates of the costs of biofuel production may be critically dependent on certain assumptions used in their calculation. As few biofuel production systems have been commercially tested there are no reliable figures for performance parameters, such as yields of fuel or feedstocks, availability and reliability of the technology and lifetimes of components, under realistic operating conditions. Variation in estimates of these values can significantly alter the economic attractiveness of a fuel production system. The economics may also be influenced by other assumptions such as the scale of the technology, the apportioning of production costs to wastes or components of dual purpose crops, and the value attributed to non-fuel products and unquantifiable benefits such as odour reduction in slurry.

Table 1
Comparative cost of conventional fuels, UK 1981

Fuel	£ GJ^{-1}	efficiency (%) of use	effective £ GJ^{-1}
Fuel oil	4.35	75–90	4.83– 5.80
Coal	2.80– 4.42	70	4.00– 6.31
Gas	2.82– 5.27	80–88	3.20– 6.99
Electricity	7.61–15.44	100	7.61–15.44
Propane	5.50	80–88	6.25– 6.88

Sources: Adapted from ADAS/BABA (1982), Department of Energy (1982), Staniforth (1982), Tougaard Pedersen (1982).

Case study 1: Straw for fuel

FEEDSTOCK PRODUCTION

Output

2.5 t straw (89% DM) per hectare per year

			$£\ ha^{-1}$	$£\ t^{-1}$
Costs				
	Variable costs	tractor fuel[1]	1.60	
		baler twine	3.30	
	Machinery depreciation and repairs			
		baler[2]	5.67	
		bale accumulator[2]	0.68	
		trailer[3]	3.30	
		tractor[4]	9.36	
	Labour[5]		13.39	
	Total		37.30	14.92

COMBUSTION

Output

475 GJ per year[6]

		$£\ y^{-1}$	$£\ GJ^{-1}$
Costs			
	Boiler – depreciation and interest[7]	500.0	
	Labour[8]	226.38	
	Total	726.38	
	Straw production costs – 49 t	731.08	
	Total	1 457.46	3.07
	Effective cost (production only)[6]		1.65

1 10 l per ha @ £0.16 per l

2 100 h per y @ 0.9 h per ha

3 200 h per y @ 3.15 h per ha

4 1000 h per y @ 4.05 h per ha

5 5.8 h per ha @ £2.31 per h

6 60% combustion efficiency

7 60 kW boiler – £2 700 (including chimney) depreciated over 12 years at 15% interest

8 2 h per t @ £2.31 per h

Sources: Calculated from Nix (1982), Tougaard Pedersen (1982)

Case study 2: Anaerobic digestion of wastes from a 4000 pig place unit

GAS PRODUCTION

Output

1.82 TJ biogas per year[1]

Costs	£ y^{-1}	£ GJ^{-1}
Long-life digester equipment[2]	4 500	
Short-life digester equipment[3]	2 245	
Maintenance	750	
Total	7 495	4.11
Effective cost (gas production only)[4]		4.90

GAS COMBUSTION

Costs	£ y^{-1}	£ GJ^{-1}
Boiler depreciation and interest[5]	568	
Maintenance	400	
Total	968	
Gas production cost	7 495	
Total	8 463	4.65

1 320 m^3 gas per day gross less 35% for digester function and losses = 208 m^3 per day net 350 days per year = 72 800 m^3 per year @ 25 MJ per m^3

2 £22 500 depreciated over 10 yrs @ 15% interest

3 £7 500 depreciated over 5 yrs @ 15% interest

4 Assuming 84% combustion efficiency

5 64 kW boiler @ £48 per kW depreciated over 12 years @ 15% interest

Sources: Adapted from ADAS/BABA (1982), Tougaard Pedersen (1982)

Case study 3: Fuels from bracken (Pteridium aquilinum)

FEEDSTOCK PRODUCTION

Output

6 tDM per hectare per year from autumn harvest yielding dry material for direct combustion or gasification to produce methanol (this system maintains bracken).
9 tDM per hectare per year from summer harvest yielding wet material for anaerobic digestion to produce biogas (this system eradicates bracken).

Costs

Item	Direct combustion	Gasification	Anaerobic digestion
	£ t^{-1}	£ t^{-1}	£ t^{-1}
Fertiliser	6.0	6.0	14.0
Cutting & collection	6.3	6.3	7.6
Densification	13.0	8.0	N/A
Storage (1 year)	2.0	6.0	32.0
Transport (20 km)	3.0	3.0	12.0
Total	30.0	29.3	65.6

FUEL PRODUCTION

Output

21 GJ combustible material per tonne DM

10.5 GJ methanol per tonne DM

9.5 GJ methane per tonne DM

Costs

Item	Direct combustion	Gasification	Anaerobic digestion
	£ GJ^{-1}	£ GJ^{-1}	£ GJ^{-1}
Conversion process	N/A	4.3	4.1
Feedstock	1.4	2.8	6.9
Total	1.4	7.1	11.0
Effective cost	2.3[1]	N/A	13.1[2]

1 Combustion efficiency of 60%

2 Combustion efficiency of 84%

Source: Adapted from Callaghan *et al* (1982)

Case study 4: Biogas from catch fuel crops

FEEDSTOCK PRODUCTION

Output

4 tDM per hectare per year

Costs

Item	£ ha^{-1}	£ t^{-1}
Fuel	6.3	
Fertiliser	55.0	
Seed	15.0	
Total	76.3	19.1

GAS PRODUCTION

Output

3 150 GJ biogas per year[1]

Costs

Item	£ y^{-1}	£ GJ^{-1}
Digestion equipment	5 600.0[2]	
Running costs	3 500.0[3]	
Total	9 100.0	
Feedstock	6 685.0[4]	
Total	15 785.0	5.0
Effective cost		6.0[5]

1 Net gas yield of 9 GJ per day operating for 350 days per year

2 200 m^3 digester costing £35 000 amortised over 20 years at 15% interest

3 10% capital cost

4 Assuming 1 tDM per day

5 Assuming 84% combustion efficiency

Sources: Calculated from Nix (1982) and data supplied in personal communications with Dr K Richards, Dr D Stafford and Mr A M M Thompson

Appendix III
Energy analysis

Energy Analysis (EA) was developed during the 1970s as a method of determining a commodity's worth by a meaningful unit of account other than that put on it by monetary cost alone' (Lewis, 1980). It involves the calculation of all the energy inputs, both direct and indirect, to a commodity to produce a Gross Energy Requirement (GER) for that commodity. The GERs of all the inputs to a production system may then be summed to produce a GER for the whole system. This is then often compared with some measure of output to derive the Net Energy (NE), or a Net Energy Ratio (NER), for the system which may sometimes be used as a basis for comparison of different systems. Energy analysis has been applied in many fields with considerable use in agriculture (Pimentel *et al,* 1973; Leach, 1976; Slesser *et al,* 1977).

Some of the more zealous proponents of EA have suggested that, in the light of future energy scarcity, it may be preferable to economics as a basis for decision-making. Many other studies have also implied that minimisation of use of finite energy resources should be the objective function in policy analysis, by their use of EA to 'rank' alternative production systems. The use of EA as a normative technique is, however, questionable. Such a single parameter theory of value suggests that energy is the most important factor in policy choice. Non-energy attributes of a production system are therefore unaccounted for, and this may lead to a misallocation of non-energy resources. Conventional economic analysis, it is argued (Webb & Pearce, 1975; Lloyd, 1980; Renborg, 1980) adequately incorporates energy considerations and EA does not provide a satisfactory alternative.

Criticism may also be levelled at other aspects of the technique; all energy sources, except labour, are aggregated, thus all energy inputs and outputs are treated solely on a per MJ basis regardless of energy quality, scarcity, or availability; a variety of techniques have also been used to account for the energy input from labour (Fluck, 1981) and a satisfactory convention has yet to be resolved. EA should be applied to the whole fuel production process, a partial analysis indicating a good energy balance for part of the process, say feedstock production, may disguise high energy costs for conversion. A proper EA for biofuel production would also have to consider the energy effects of the changes arising from a large-scale biofuel production programme: what is the energetic cost of importing more food or intensifying production on reduced acreage?

Case study 1: Straw for fuel

	£ y^{-1}	MJ £$^{-1}$	Total GJ y^{-1}
Baler	111.13	30	3.3
Bale accumulator	13.33	35	0.5
Trailer	64.68	35	2.3
Tractor	183.46	22	4.0
Tractor fuel	31.36	271	8.5
Baler twine	64.68	41	2.7
Boiler	500.00	50	25.0

Total support energy input (excluding labour) = 46.3 GJ per y

Labour energy requirement = 58 MJ per y

Gross energy requirement = 58.6 GJ per y

Useful energy output = 475 GJ per y

Net energy output = 416.4 GJ per y

Energy ratio = 8:1

Case study 2: Anaerobic digestion of pig wastes

	£ y^{-1}	MJ £$^{-1}$	Total GJ y^{-1}
Long-life equipment	4 500	50.0	225
Short-life equipment	2 245	36.4	82
Maintenance	750	36.4	27
Boiler	568	50.0	28
Maintenance	400	36.4	15

Total support energy requirement = 377 GJ per y

Total energy output = 1 820 GJ per y

Net energy output = 1 443 GJ per y

Energy ratio = 5:1

Note: 1974 data deflated to 1980 values

Sources: Calculated from Wright (1975), CSO (1978, 1982), Doering (1980), Fluck (1981), Massey Ferguson (1981, 1982), ADAS/BABA (1982), Catling (1982), Common & McPherson (1982), Nix (1982) from data in Appendix II and using data supplied in personal communications by B J Anslow & G M Pocket.

Appendix IV
Directory of UK energy from biomass organisations

The following organisations are known to have an interest in some aspect, or aspects, of biofuel production or utilisation. Interests range from information gathering and research, development and demonstration, to equipment manufacture and sales, and technical services and consultancy. Listing of commercial organisations should not be taken, necessarily, as a recommendation.

FEEDSTOCK PRODUCTION

AGRICULTURAL RESIDUES

S B C Larkin
National College of Agricultural Engineering, Wrest Park, Silsoe, Beds
Research: assessment of availability of agricultural residues

R M Morris
The Open University, Biosystems Group, Systems Discipline, Walton Hall, Milton Keynes
Research: assessment of availability of agricultural residues

AQUATIC/MARINE

T J Holt, J M Kain
University of Liverpool, Department of Marine Biology, Port Erin, Isle of Man
Research: offshore seaweed as a biofuel source

J G Morley
University of Nottingham, Wolfson Institute of Interfacial Technology, University Park, Nottingham
Research: offshore seaweed as a biofuel source

NATURAL VEGETATION

T V Callaghan, G J Lawson, R Scott
Institute of Terrestrial Ecology, Merlewood Research Station, Grange-over-Sands, Cumbria
Research: natural vegetation and high-yielding introduced species

ENERGY CROPS

S Cousins, S Newman, S Smith
The Open University, Energy Research Group, Walton Hall, Milton Keynes
Research: mixed cropping for food and fuel; wetland species for energy

S P Carruthers
University of Reading, Department of Agriculture and Horticulture, Earley Gate, Reading, Berks
Research: catch crops for fuel

FORESTRY

I R Brown, C P Mitchell
University of Aberdeen, Forestry Department, St Machar Drive, Aberdeen
Research and development: short rotation forestry for energy; improvement of Scottish birches

K G Stott
University of Bristol, Long Ashton Research Station, Long Ashton, Bristol, Avon
Research: coppiced trees

M Lofthouse
Forestry Commission, Milyard, Smithton, Inverness
Research and development: whole-tree chipping

G Tuley
Forestry Commission, Forest Research Station, Alice Holt Lodge, Wrecclesham, Farnham, Surrey
Research: exotic broadleaved trees in Britain

M L Pearce
Forestry Commission, Research and Development Division, Westonbirt Arboretum, Tetbury, Gloucs
Research: coppiced trees

H D McLain
Loughry College of Agriculture and Food Technology, Cookstown, County Tyrone, Northern Ireland
Research and development: coppiced trees production and harvesting

P J Newbould, C Mulholland
New University of Ulster, Department of Environmental Science, Coleraine, County Londonderry, Northern Ireland
Research: biomass production from short-rotation coppice

LAND RESOURCES

P Downing, O Brandon
Dartington Amenity Research Trust, Shinners Bridge, Dartington, Totnes, Devon
Research: availability of land in the UK for wood energy plantations

R G H Bunce, C J Barr, H A Whittaker
Institute of Terrestrial Ecology, Merlewood Research Station, Grange-over-Sands, Cumbria
Research: availability of land in the UK for wood energy plantations

A M M Thompson, R B Tranter
University of Reading, Centre for Agricultural Strategy, 1 Earley Gate, Reading, Berks
Research: availability of land in the UK for wood energy plantations

M Chadwick
University of York, Derelict Land Reclamation Research Unit, Department of Biology, Heslington, York
Research: assessment of industrial waste land for growth of energy crops

CONVERSION PROCESSES

ANAEROBIC DIGESTION

V C Nielsen
ADAS Farm Waste Unit, Government Buildings, Coley Park, Reading, Berks
Research: monitoring of on-farm anaerobic digestion

C Davies
ADBAC, The Great Yard, Marridge Hill, Ramsbury, Marlborough, Wilts
System design: feasibility study and design and build services for effluent treatment and anaerobic digestion

Agricultural Engineers Association (AEA), 6 Buckingham Gate, London SW1E 6VU
Trade association: interests include anaerobic digestion

Ames Crosta Babcock Ltd, Gregge Street, Heywood, Lancs
System design, equipment manufacture and sales: sludge digestion systems

B H Kirsop
Agricultural Research Council, Food Research Institute, Colney Lane, Norwich, Norfolk
Research: anaerobic digestion

N C Worth
Bio-digestion Services Ltd, The Business Centre, Claughton Road, Birkenhead, Merseyside
Equipment manufacture and sales, consultancy: anaerobic digestion equipment

S D Pannel
Bio-Energy International Ltd, Sandy Lane Industrial Estate, Stourport-on-Severn, Worcs
System design equipment manufacture and sales, consultancy: waste treatment equipment

S Whittaker
Biomass International, PO Box 5, Milnthorpe, Cumbria
System design, equipment manufacture and sales: plant for the production of energy from liquid and solid waste

Biomechanics Ltd, Caxton House, Wellesley Road, Ashford, Kent
Research, development and demonstration, system design, technical services and consultancy: anaerobic treatment

P J Meynell
British Anaerobic and Biomass Association Ltd, The White House, Little Bedwyn, Marlborough, Wilts
Trade association: anaerobic digestion and general energy from biomass

British Water and Effluent Treatment Plant Association (BWETPA), 51 Castle Street, High Wycombe, Bucks

Trade association: interests include anaerobic digestion

D A Stafford, S P Etheridge
University College Cardiff, Department of Microbiology, AD Research Unit, Cleppa Park Research Station Coedkernew, Nr Newport, Gwent

Research and development: anaerobic digestion

N Wates
Commonwork Enterprises Ltd, Bore Place, Chiddingstone, Edenbridge, Kent

Research, development and demonstration: farm operating digester on waste from 300 head dairy herd

Cooper, MacDonald and Partners, Loxford House, East Street, Epsom, Surrey

Consultancy: civil and structural engineering including sewage treatment

R Tillier
Degremont Laing Ltd, Aquazur House, Elstree Way, Borehamwood, Herts

System design, equipment manufacture and sales: anaerobic digestion equipment

S Burgess
Dunlop Bioprocesses Ltd, Kingsbury Road, Erdington, Birmingham

Equipment manufacture and sales, consultancy: anaerobic digestion and fermentation/distillation

Energy and Waste Systems Ltd, Bridge House, Westbury, Wilts

Equipment manufacture and sales: plant for anaerobic digestion

M Chesshire
Farm Gas Ltd, Industrial Estate, Bishops Castle, Salop

Equipment manufacture and sales: anaerobic digesters

Farrow Irrigation Ltd, Horseshoe Road, Spalding, Lincs

System design, equipment manufacture and sales: liquid waste handling equipment

S Wray
Hamworthy Engineering Limited, Fleets Corner, Poole, Dorset

Equipment manufacture and sales: supply of anaerobic digestion plant

M Sibley
A O Smith Harvestore Products Ltd, Eye, Suffolk
Equipment manufacture and sales: anaerobic digestion

D Campbell, J F Rees
Harwell Environmental Safety Group, AERE, Harwell, Oxon
Consultancy: waste disposal and treatment including anaerobic digestion

J E Marshall
IMI Energy Systems, PO Box 216, Witton, Birmingham
System design, equipment manufacture and sales: waste treatment processing and handling; supply of treated refuse to industry

Jones and Attwood Limited, Titan Works, Stourbridge, West Midlands
Equipment manufacture and sales: waste treatment systems

A Cheyney
London Brick Company, Stewartby, Bedford
Research, development and demonstration: methane from landfill

S J Pirt
University of London, Queen Elizabeth College, Department of Microbiology, Campden Hill Road, London W8 7AH
Research: anaerobic digestion; energy production from algae

J D Bu'Lock
University of Manchester, Department of Chemistry, Oxford Road, Manchester
Research: anaerobic digestion

G H Fynn
University of Manchester Institute of Science and Technology, Department of Biochemistry, PO Box 88, Manchester
Research: biochemistry of methane production

B Atkinson, D E Brown
University of Manchester Institute of Science and Technology, Department of Chemical Engineering, PO Box 88, Manchester
Research: fluidised bed waste treatment; cellulase synthesis

B F Pain
National Institute for Research in Dairying (NIRD), Shinfield, Reading, Berks
Research and development: anaerobic digestion

A D McKelvie
North of Scotland College of Agriculture, 581 King Street, Aberdeen
Research and development: treatment of cattle and pig slurry by anaerobic digestion

P N Hobson
Rowett Research Institute, Bucksburn, Aberdeen
Research and development: anaerobic digestion of animal wastes

E Stobart
SCS Biotechnology Ltd, Elmsfield, Chipping Norton, Oxon
Consultancy: anaerobic digestion and liquid separation

A L Best
SATEC Ltd, PO Box 12, Weston Road, Crewe, West Midlands
Equipment manufacture and sales, consultancy: anaerobic digestion, fermentation/distillation and single cell protein from waste

G F Shattock
5 Orchard Close, Long Lane, Tilehurst, Reading, Berks
Consultancy: agricultural waste treatment

N R McFarlane
Shell International Petroleum Company, Non-traditional Business Division, Shell Centre, London
Research: anaerobic digestion fermentation/distillation and single cell protein from waste

M Provan
University of Strathclyde, Department of Applied Microbiology, Royal College Building, 204 George Street, Glasgow
Research: energy from parkland wastes, anaerobic digestion of grass clippings

F R Hawkes, J R S Floyd
Polytechnic of Wales, Department of Science, Pontypridd, Mid-Glamorgan
Research: anaerobic digestion

J Cuthbert
Water Research Centre, Stevenage Laboratory, Elder Way, Stevenage, Herts
Research and technical services: anaerobic waste treatment

FERMENTATION

Alcon Biotechnology Ltd, 1 Buckingham Street, Portsmouth

Research: continuous fermentation

H J Grout, M English
W S Atkins Group Ltd, Woodcote Grove, Ashley Road, Epsom, Surrey

Research: continuous integrated fermentation and distillation

A N Emery
University of Birmingham, Department of Chemical Engineering, PO Box 363, Birmingham

Research: fermentation

Brewing Research Foundation, Lyttel Hall, Nuffield, Redhill, Surrey

Research: fermentation

D Essien, D L Pyle
Imperial College, Department of Chemical Engineering and Technology, London SW7 2BY

Research: fermentation

A Porteous
The Open University, Department of Engineering Mechanics, Walton Hall, Milton Keynes

Research: acid hydrolysis of cellulosic feedstocks

R Greenshields
University College of Swansea, Biotechnology Centre, Singleton Park, Swansea

Research: fermentation process and fermentor design

THERMAL PROCESSING

A V Bridgewater, B N Hatt, R N Shand, E L Smith
University of Aston in Birmingham, Department of Chemical Engineering, Gosta Green, Birmingham

Research, development and demonstration: thermal processing

Constructors John Brown Ltd, CJB House, Buckingham Street, Portsmouth

Research, development and demonstration: gasification and methanol production

Tropical Products Institute, Industrial Development Department, Culham, Abingdon, Oxon

Research, development and demonstration: gasifier/engine system based on charcoal for motive power in developing countries; biogas in developing countries

C J V Denning, R Bowers, R J McLellan
Wellman Mechanical Engineering, Gasification Division, Cornwall Road, Smethwick, West Midlands

Research, development and demonstration: gasification and methanol production

FUEL UTILISATION

SOLID FUEL COMBUSTION

Air Plant (Sales) Ltd, 295 Aylestone Road, Leicester

Equipment manufacture and sales: waste-fired boiler, orthodox incinerator, pyrolytic incinerator, woodburning equipment, underfed stoker

Allied Boilers Ltd, Unit 7, Belgrave Mills, Honey Lane, Oldham, Lancs

Equipment manufacture and sales: fluidised bed boiler, waste heat recovery boilers

Babcock Power Ltd, 165 Gt Dover Street, London

Equipment manufacture and sales: waste-fired boiler, fluidised bed boiler, waste heat recovery boiler, orthodox incinerator

Bayford Tomlinson Ltd, 6 Lidgett Lane, Leeds

Equipment manufacture and sales: waste-fired boiler

B & E Boilers, Castlehampstead Road, Bracknell, Berks

Equipment manufacture and sales: woodburning equipment

Beverley Engineering Ltd, Billingshurst, West Sussex

Equipment manufacture and sales: pyrolytic incinerator

Blue Circle Industries plc, Blue Circle Technical Department, New Road, Gravesend, Kent

Research, development and demonstration: systems for using town refuse, waste tyres, landfill and straw as fuel for cement works

Chiltern Timber Heating, Greenstead, Magpie Lane, Coleshill, Amersham, Bucks
Equipment manufacture and sales: domestic woodburning equipment

Clarke Chapman, Derby Works, Sinfin Lane, Derby
Equipment manufacture and sales: woodburning equipment, suspension-fired

Cleerburn Ltd, Northway, Walworth Industrial Estate, Andover, Hants
Equipment manufacture and sales: orthodox incinerator, cyclone incinerator

Combat Ltd, Oxford Street, Bilston, West Midlands
Equipment manufacture and sales: waste-fired boiler, waste heat recovery boiler

Combustions Ltd, Cox Lane, Chessington, Surrey
Equipment manufacture and sales: domestic woodburning equipment

Cormall Ltd, Evanton Place, Thornliebank Industrial Estate, Glasgow
Equipment manufacture and sales: machinery for processing straw

Danks of Netherton Ltd, 257 Halesown Road, Dudley, West Midlands
Equipment manufacture and sales: waste-fired boiler, fluidised bed boiler, waste heat recovery boiler, woodburning equipment, suspension fired

Deborah Fluidised Combustion Ltd, 8 Davey Drive, NW Industrial Estate, Peterlee, County Durham
Equipment manufacture and sales: fluidised bed boiler, cyclone incinerator

Dowson and Mason Ltd, Alma Works, Levenshulme, Manchester
Equipment manufacture and sales: fluidised bed boiler, pyrolytic incinerator

Dragon Straw Burners, Sibsey, Boston, Lincs
Equipment manufacture and sales: straw-burning space heaters

Dust Control Components, 27 Crendon Street, High Wycombe, Bucks
Equipment manufacture and sales: suspension-fired woodburning equipment

Energy Equipment Co Ltd, Energy House, Hockliffe Street, Leighton Buzzard, Beds
Equipment manufacture and sales: woodburning equipment, fluidised bed boiler

Exenco Ltd, Holder Road, Aldershot, Hants
Equipment manufacture and sales: straw-burning hot water boiler

Extracta Engineering Ltd, Holder Road, Aldershot, Hants
Equipment manufacture and sales: waste-fired boiler, underfed stoker, woodburning equipment

P Teisen
Farm 2000 Ltd, Bell Lane, Studley, Warwicks
Equipment manufacture and sales: straw and woodburning boilers and heating systems

Flameless Furnaces Ltd, Throgmorton House, 15 Copthall Avenue, London EC2R 7BU
Equipment manufacture and sales: fluidised bed boiler

Foster Wheeler Power Products Ltd, PO Box 160, Greater London House, Hampstead Road, London NW1 7QN
Research, development and demonstration: gasification of biomass.
Equipment manufacture and sales: waste-fired boiler, fluidised bed boiler, waste heat recovery boiler

General Engineering Radcliffe (1979) Ltd, Bury Road, Radcliffe, Manchester
Equipment manufacture and sales: rotary chamber incinerator

Gerard Hare (Engineers) Ltd, Rockborne Mews, 65a Rockbourne Road, Forest Hill, London SE23 2DE
Equipment manufacture and sales: woodburning equipment, cyclone incinerator

Hoval Farrar Ltd, Northgate, Newark, Notts
Equipment manufacture and sales: pyrolytic incinerator, waste heat recovery boiler

The Incinerator Company Ltd, Howard Road, Eaton Socon, St Neots, Huntingdon, Cambs
Equipment manufacture and sales: orthodox incinerator

Incol Incineration (Unit Eng Co Ltd), 32 Market Street, Altrincham, Cheshire
Equipment manufacture and sales: orthodox incinerator, waste heat recovery boiler

Robert Jenkins Systems Ltd, Wortley Road, Rotherham, South Yorks
Equipment manufacture and sales: pyrolytic incinerator

Lucas Furnaces Ltd, Western Way, Wednesbury, Staffs

Equipment manufacture and sales: woodburning equipment, cyclone incinerator

Lyon and Pye Ltd, 543 Prescot Road, St Helens, Merseyside

Equipment manufacture and sales: orthodox incinerator, pyrolytic incinerator

Motherwell Bridge Tacol Ltd, 64/76 High Street, Croydon, Surrey

Equipment manufacture and sales: orthodox incinerator

NEI Ltd, Sinfin Lane, Derby

Equipment manufacture and sales: waste-fired boiler, fluidised bed boiler, waste heat recovery boiler

New Air Technical Services Ltd, 34 Bell Street, Wigston Magna, Leicester

Equipment manufacture and sales: waste-fired boiler, orthodox incinerator, pyrolytic incinerator

W E Klinner, A C Knight
National Institute of Agricultural Engineering (NIAE), Wrest Park, Silsoe, Bedford

Research and development: mechanical processing of straw for fuel

B J Wilton
University of Nottingham, School of Agriculture, Sutton Bonington, Loughborough, Leics

Research and development: fuel utilisation, direct combustion; whole-crop harvesting

Parkinson Cowan GWB, PO Box 4, Burton Works, Dudley, West Midlands

Equipment manufacture and sales: woodburning equipment, suspension firing

Robeys of Lincoln Ltd, PO Box 23, Caswick Road, Lincs

Equipment manufacture and sales: waste heat recovery boiler

Savonheat Ltd, Denmark House, 6-8 College Street, Petersfield, Hants

Equipment sales: central heating boilers

Scanfield Boilers Ltd, Capel House, Harn Street, Ashford, Kent

Equipment manufacture and sales: straw-fired boilers

SIP Woodprocessing Limited, Combe Sydenham, Monksilver, Somerset

Equipment manufacture and sales: woodburning equipment, fluidised bed boiler

Stoneplatt Fluidfire Ltd, Unit 11, Washington Street, Netherton, Dudley, West Midlands

Equipment manufacture and sales: woodburning equipment, suspension-fired

Stoneplatt Fluidfire Ltd, 56 Second Avenue, Pensnott Trading Estate, Brierley Hill, West Midlands

Equipment manufacture and sales: fluidised bed boiler

Thorn EMI Boilers Limited (Parkinson Cowan GWB), PO Box 4, Burton Works, Dudley, West Midlands

Equipment manufacture and sales: waste-fired boiler, fluidised bed boiler

Towler & Son Ltd, Telford Road, Gorse Lane Industrial Estate, Clacton-on-Sea, Essex

Equipment manufacture and sales: waste-fired boiler, pyrolytic incinerator, wood burning, fixed grate water tube

Universal Machinery and Services Ltd, Millshaw, Leeds

Equipment manufacture and sales: orthodox incinerator

Woodburners Association of Retailers and Manufacturers (WARM), PO Box 35, Stoke-on-Trent, Staffs

Trade association: woodburners

LIQUID FUELS

V Moore
Berkshire College of Agriculture, Burchetts Green, Maidenhead, Berks

Research and development: vegetable oils in diesel engines

British Petroleum, BP Research Centre, Sunbury

Research and development: use of supplementary fuels in diesel engines, fuel blends include 10% methanol, ethanol, vegetable oils and vegetable oil esters

E M Goodger
Cranfield Institute of Technology, School of Mechanical Engineering, Cranfield, Bedford

Research: availability, properties and performance of alternative fuels

Lucas CAV Ltd, PO Box 36, Warple Way, London W3 7SS

Research: engine tests on vegetable oils and added ethanol

R W Radley
National College of Agricultural Engineering, Cranfield Institute of Technology, Silsoe, Bedford
Research: alternative fuels for diesel engines

Perkins Engines Ltd, Eastfield, Peterborough, Cambs
Research: diesel engine tests using vegetable oils and blends, tallow, mono-esters, alcohol blends and emulsions and additivated alcohols

K Morgan
University of Reading, Department of Agriculture and Horticulture, Earley Gate, Reading, Berks
Research: engine modification for vegetable oils, multi-fuel engines, stirling engine

Ricardo Consulting Engineers plc, Bridge Works, Shoreham-by-Sea, West Sussex
Research: diesel engines tested with methanol, diesel and vegetable oils

GASEOUS FUELS

J W Dodsworth
British Leyland, Lode Lane, Meteor Works, Solihull, West Midlands
Research: running engines on gas

B McCarthy
Bristol Pneumatic Ltd, Lodge Causeway, Fishponds, Bristol
Equipment manufacture and sales: design and manufacture of air and gas compressors

D Jones
GEC Dorman Diesels Ltd, Tixall Road, Stafford
Equipment manufacture and sales: engines for biofuels/electricity generation

D J Picken
Leicester Polytechnic, PO Box 143, Leicester
Research: biogas in internal combustion engines

R West
National Institute of Agricultural Engineering (NIAE), Wrest Park, Silsoe, Bedford
Research and development: biogas in engines

R W Kelcher
SERCK Heat Transfer, Warwick Road, Birmingham

Equipment manufacture and sales: supply and installation of total energy systems for production of heat and electricity

C J Hewitt
Reading Industrial Engines, Stadium Way, Reading, Berks

Development, demonstration and equipment sales: power generation from engines running on biogas

IMPLEMENTATION AND POLICY

H Rothman
University of Aston in Birmingham, Technology Policy Unit, Gosta Green, Birmingham

Research: socio-economic and environmental impacts of new technology

M B Quiggin
University of Bradford, Postgraduate School of Studies in Science and Society, Bradford, West Yorks

Research: evaluation of energy supply strategies

P J Gummett
University of Manchester, Department of Liberal Studies in Science, Faculty of Science, Oxford Road, Manchester

Research: policy for alternative energy sources

S P Carruthers, B G F Weitz
University of Reading, Centre for Agricultural Strategy, 1 Earley Gate, Reading, Berks

Research: energy in agriculture; biofuel production, implementation and policy

P D Dunn
University of Reading, Department of Engineering, Whiteknights, Reading, Berks

Research, development and demonstration: implementation of biomass systems

C Lewis, M Slesser
University of Strathclyde, Energy Studies Unit, Glasgow, Strathclyde

Research: biofuel implementation and policy; energy analysis

R Harrison
Sunderland Polytechnic, Division of Physics, Edinburgh Building, Chester Road, Sunderland, Tyne and Wear

Research: assessment of alternative energy sources

R J Unsworth
University of Sussex, School of Mathematical and Physical Sciences, Physics Building, Falmer, Brighton, West Sussex

Research: assessment of alternative energy sources

J H Cheshire, D Scott-Kemmis, A J Surrey
University of Sussex, Science Policy Research Unit, Falmer, Brighton, West Sussex

Research: science, research and energy policy

GENERAL

J Coombs, D O Hall
Bio-Services, King's College, 68 Half Moon Lane, London SE24 1JF

Consultancy and information service: energy from biomass

L G Plaskett
Biotechnical Processes Ltd, Hillsborough House, Ashley, Tiverton, Devon

Consultancy: biofuel production and utilisation

G H King
Energy Technology Support Unit, AERE, Harwell, Didcot, Oxon

Research development and demonstration: coordination of UK fuels from biological materials research programme

C E Dodson, J Keable
Helix Multiprofessional Services, Mortimer Hall, Mortimer, Reading, Berks

Consultancy: particularly for anaerobic digestion

G F McRobie
Intermediate Technology Development Group (ITDG), 9 King Street, London WC2

Research and consultancy: intermediate technology, including fuel production, particularly for developing countries

G Leach
International Institute for Environment and Development, 10 Percy Street, London W1P 0DR

Research and information: energy supply including biofuels; low energy futures

Solar Trade Association Ltd, 26 Store Street, London WC7E 7BT

Trade association: solar energy including biofuels

United Kingdom International Solar Energy Society (UKISES), 19 Albemarle Street, London

Society: all aspects of solar energy including biofuels

G Ader
WDH Services, Acorn House, High Street, London SE2O 7DS

Consultancy: biofuel production and utilisation

Appendix V
Glossary

Aerobic decomposition The biological breakdown of biomass in the presence of air; commonly used as a treatment for animal wastes.

Anaerobic digestion (AD) The biological breakdown of biomass in the absence of air to produce a methane-rich 'biogas'; used in waste treatment.

Anaerobic filter A type of anaerobic digester in which the feedstock flows through a high surface area filter-bed in which a large population of methanogenic micro-organisms is immobilised. Designed for use with liquid wastes.

Bagasse The remains of sugar cane after the juice has been extracted.

Beer The dilute (5–12%) alcohol solution produced by fermentation.

Bioclimatic region A region possessing distinctive biological and climatic characteristics.

Biogas A mixture of mainly methane (CH_4), and carbon dioxide (CO_2) produced by AD.

Biochemical Oxygen Demand (BOD) A measure of the oxygen requirement to biologically oxidise a material.

Break crop A full-season crop grown on land normally under continuous cultivation of another crop, typically cereals, to avoid build-up of disease and pests, and to restore soil fertility.

Carboxylolysis Breakdown of biomass by treatment with steam and CO in the presence of a sodium carbonate catalyst.

Catch crop A crop grown on land available between the harvest of one crop and the planting of the next.

Cell recycle The recovery and return of microbial cells washed out of a continuous fermentation reactor.

Cell washout The loss of microbial cells from a continuous fermentation reactor, eg yeast in brewing.

Chemical Oxygen Demand (COD) A measure of the oxygen requirement to chemically oxidise a material in solution.

Char The solid product, largely carbon, of pyrolysis (*cf* charcoal).

Cryophilic digestion Anaerobic digestion at low (eg 15° C) temperatures.

Direct liquefaction The chemical conversion of solid biomass directly to liquids.

Efficiency The ratio of outputs to inputs of a process in terms of the stated parameter (eg energy).

Ethanol A colourless liquid (C_2H_5OH), used in beverages, commonly known as 'alcohol'.

Exotic A species of plant or animal not native to a country.

Feed Animal fodder.

Feedstock Biomass used as the raw material for a conversion process.

Fluidised bed A bed of inert material kept in a fluid state by gas or liquid passed through.

Food Products for human nutrition.

Freeze crystallisation The separation of a mixture of two liquids by freezing.

Gasification The thermal processing of biomass to produce mainly gases.

Gas scrubber A device to remove impurities from gases.

Hardware Machinery or equipment.

Hydraulic digester A type of anaerobic digester which has hydraulic regulation of the flow through the system. Solids are retained whilst liquids pass quickly through.

Hydrodynamic analogue A model which uses the flow of water as an analogy for other flows in the system.

Hydrogenation The addition of hydrogen to chemical compounds.

Hydrolysis The breakdown of chemical compounds by water.

Lignin A complex organic polymer found in wood, not readily susceptible to biological degradation, often combined with cellulose.

Macrophyte A non-microscopic plant.

Mesophilic digestion Anaerobic digestion at medium temperature range (30–40°C).

Methanogenic Producing methane (CH_4).

Methanol A colourless toxic liquid (CH_3OH), commonly known as 'wood alcohol'.

'MINO' Process A gasification process developed at the Royal Institute of Technology, Stockholm, Sweden, which can handle wet feedstocks (up to 50% MC).

Photo-biology The study of the effect of light on biological systems, particularly photosynthesis.

Photo-chemistry The study of the effect of light on chemical reactions.

Photoelectro-chemistry The study of chemical systems which liberate electrons under light.

Plantation An area of land used for the production of a single, usually tree, crop.

Plug-flow digester A type of anaerobic digester in which no mixing occurs. High solids feedstocks move through the length of the digester as a result of horizontal displacement by influent feedstocks.

Poikilotherms Animals which do not endogenously regulate their body temperature ('cold-blooded animals').

Pyrolysis Destructive distillation in the absence of air.

Renewables Renewable sources of energy, such as biomass, solar, and wind.

Retention time The length of time a feedstock remains in a digester.

Solvent extraction The separation of two liquids by differential solution in a solvent, or of liquid from a solid.

Stover The stalk of maize left after the cobs have been harvested.

Substitute natural gas (SNG) Methane (CH_4).

Synthesis gas A mixture of carbon monoxide (CO) and hydrogen (H_2), produced in the synthesis of methanol.

Thermophilic digestion Anaerobic digestion at high ($>50^{o}C$) temperatures.

References

ADAS/BABA (1982) *The economics of anaerobic digesters: a discussion document of an ADAS/BABA working party.* London: Ministry of Agriculture, Fisheries and Food.

Ader, G & Buck, F R (1979) *Organic wastes as an energy source.* Report on a study conducted on behalf of the Department of Energy, Energy Technology Support Unit. West Wickham, Kent: Ader Associates.

American Society of Agricultural Engineers (1982) *Vegetable oil fuels.* ASAE Publication, 4–82. Michigan: American Society of Agricultural Engineers.

Balls, R C (1982). Straw for fuel in horticulture – a case study of a nursery in Lincolnshire during the 1982 season. In: Staniforth, A R (Ed) *Report on 6th straw utilisation conference.* Oxford: Agricultural Development and Advisory Service/National Farmers Union.

Bandel, W & Heinrich, W (1983) Vegetable oil derived fuels and problems related to their use in diesel engines. In: Strub, A, Chartier, P & Schleser, B (Eds) *Energy from biomass,* 822–829. London: Applied Science Publishers.

Barry, M, Colleran, E & Wilkie, A (1982) Two-stage anaerobic digestion of organic residues and energy crops. In: Vogt, F (Ed) *Energy conservation and use of renewable energies in the bio-industries 2,* 75–87. Oxford: Pergamon Press.

Bather, D M & Carruthers, S P (1980) Energry from agriculture: catch crops as a potential fuel source in the UK. In: Vogt, F (Ed) *Energy conservation and use of renewable energies in the bio-industries,* 9–22. Oxford: Pergamon Press.

Beldman, B, Voragen, A G J, Rombouts, F M & Pilnik, W (1982) Liquefaction and saccharification of agricultural biomass. In: Grassi, G & Palz, W (Eds) *Energy from biomass,* 266–272. Dordrecht, Holland: D Reidel Publishing Company.

Bergman, K G (1981) Potential for energy cropping in Swedish agriculture. In: Palz, W, Chartier, P & Hall, D O (Eds) *Energy from biomass,* 896–902. London: Applied Science Publishers.

Bernhardt, W (1981) Alternative fuels from biomass and their use in transport. In: Palz, W, Chartier, P & Hall, D O (Eds) *Energy from biomass,* 815–825. London: Applied Science Publishers.

Bishop, C F H (1980) Danish experience of straw as a fuel. In: Staniforth, A R (Ed) *Report on 'straw as a fuel' conference.* Oxford: Ministry of Agriculture, Fisheries and Food.

Bliss, C & Blake, D O (1977) *Silvicultural biomass farms volume V: conversion processes and costs.* MITRE Technical report number 7347.

Blomkvist, G, Liinanki, L, Lindman, N & Sjoberg, S O (1983) Swedish gasification research and process development at RIT. In: Strub, A, Chartier, P & Schleser, G (Eds) *Energy from biomass,* 896–900. London: Applied Science Publishers.

Brandon, O H, Barr, C, Bunce, R G H, Downing, P, Mitchell, P, Pearce, M L, Tranter, R B & Whittaker, H A (1983) *Land availability for wood energy plantations.* Technical report to the Department of Energy, Energy Technology Support Unit. Harwell: Energy Technology Support Unit.

Buckland, R & Buick, C (1980) On farm production of alternative fuels: some basic issues. *Quarterly Review of the Rural Economy,* **2** (2), 186–191.

Bühner, T & Kögl, H (1981). Possibilities and limits of using biomass as substitutes for exhaustible resources – A systems analysis approach to producing fuels from rape seed in the Federal Republic of Germany. In: Palz, W, Chartier, P & Hall, D O (Eds) *Energy from biomass,* 854–862. London: Applied Science Publishers.

Bungay, H R (1981) *Energy the biomass options.* New York: Wiley.

Burton, A, De Zutter, D, Grange, P, Poncelet, G & Delmon, B (1983) Catalytic liquefaction of wood material. In: Strub, A, Chartier, P & Schleser, G (Eds) *Energy from biomass,* 935–939. London: Applied Science Publishers.

CAS (1980a) *Strategy for the UK forest industry.* CAS Report 6. Reading: Centre for Agricultural Strategy.

CAS (1980b) *The efficiency of British agriculture.* CAS Report 7. Reading: Centre for Agricultural Strategy.

CEC (1968) *Memorandum on the reform of agriculture in the European Economic Community.* COM (68) 1000. Brussels: Commission of the European Communities.

CSO (1978) *Annual abstract of statistics 1978.* London: HMSO.

CSO (1982) *Annual abstract of statistics 1982.* London: HMSO.

Callaghan, T V, Millar, A, Powell, D & Lawson, G J (1978) *Carbon as a renewable resource in the UK – conceptual approach.* Report to the Department of Energy, Energy Technology Support Unit. Grange-over-Sands: Institute of Terrestrial Ecology.

Callaghan, T V, Scott, R & Whittaker, H A (1981) *The yield development and chemical composition of some fast-growing indigenous and naturalised British plant species in relation to management as energy crops.* Report to the Department of Energy, Energy Technology Support Unit. Grange-over-sands: Institute of Terrestrial Ecology.

Callaghan, T V, Scott, R & Lawson, G J (1982) Biofuel production from natural vegetation in Great Britain. In: Grassi, G & Palz, W (Eds) *Energy from biomass,* 30–36. Dordrecht, Holland: D Reidel Publishing Company.

Carruthers, S P (1981a) The productivity of catch crops grown for fuel. In: Palz, W, Chartier, P & Hall, D O (Eds) *Energy from biomass,* 97–102. London: Applied Science Publishers.

Carruthers, S P (1981b) Studies on the yield of catch crops grown for fuel. In: Chartier, P & Palz, W (Eds) *Energy from biomass,* 20–24. Dordrecht, Holland: D Reidel Publishing Company.

Carruthers, S P (1982) Growing catch crops for fuel – productivity, costs and implementation in the UK. In: Grassi, G & Palz, W (Eds) *Energy from biomass,* 37–42. Dordrecht, Holland: D Reidel Publishing Company.

Carruthers, S P (1983) Catch crops for energy. In: Strub, A, Chartier, P & Schleser, G (Eds) *Energy from biomass,* 127–131. London: Applied Science Publishers.

Catling, H (1982) (Ed) *Green book of tractors, farm and forestry equipment.* Redhill: Industrial Newspapers Ltd.

Chrysostome, G & Lemasle, J M (1983) Pressurised oxygen blown fluidised bed wood gasifier. In: Strub, A, Chartier, P & Schleser, G (Eds) *Energy from biomass,* 402–405. London: Applied Science Publishers.

Common, M S & McPherson, P (1982) A note on energy requirements calculations using the 1968 and 1974 UK input-output tables, *Energy Policy,* **10,** 42–48.

Costa Ribeiro, C & Castello Branco, J R (1981) Stillage: a resource disguised as a nuisance. *Process Biochemistry,* **16** (3), 8–13.

DART (1983) *Small woods on farms.* CCP 143. Cheltenham: Countryside Commission.

Demuynck, M G, Naveau, H P & Nyns, E J (1983) Assessment study on biogas plants in Europe. In: Strub, A, Chartier, P & Schleser, G (Eds) *Energy from biomass,* 620–625. London: Applied Science Publishers.

Dennington, V N & Chadwick, M J (1979) *An assessment of the potential of derelict and industrial wasteland for the growth of energy crops.* Report to the Department of Energy, Energy Technology Support Unit. York: University of York.

Dennington, V N & Chadwick, M J (1981) Forest energy crops from derelict and waste land. In: Palz, W, Chartier, P, Hall, D O (Eds) *Energy from biomass,* 251–256. London: Applied Science Publishers.

Department of Energy (1979) *Energy technologies for the United Kingdom.* Energy paper number 39. London: HMSO.

Department of Energy (1982) *Digest of UK energy statistics.* London: HMSO.

Dermott, W (1978) Achieving maximum yields in practice, likely developments and major constraints. In: MAFF *Maximising yields of crops.* London: HMSO.

Dodd, R B & Roberts, D L (1981) *On-farm ethanol pilot plant design and operation.* Paper number 81–3568 presented at a meeting of the American Society of Agricultural Engineers, Chicago, December, 15–18.

Doering, O C (1980) Accounting for energy in farm machinery and buildings. In: Pimentel, D (Ed) *Handbook of energy utilisation in agriculture,* 9–14. Boca Raton, Florida: CRC Press Inc.

Dohne, E (1981) Generation of current from biogas. In: Palz, W, Chartier, P & Hall, D O (Eds) *Energy from biomass,* 919–921. London: Applied Science Publishers.

Donellan, T & Newall, P J (1982) The biotechnology and economics of full-scale anaerobic upflow fixed bed reactors. In: Vogt, F (Ed) *Energy conservation and use of renewable energies in the bio-industries 2,* 262. Oxford: Pergamon Press.

Dunn, P D & Haskew, D W (1979) *Fuels from biological materials: production-conversion matching.* Report to the Department of Energy, Energy Technology Support Unit. Reading: University of Reading.

Emery, A N & Kent, C A (1979) Fuel alcohol production from biomass. In: *Energy from the biomass.* Watt Committee on Energy, Report number 5. London: Watt Committee on Energy Ltd.

English, B C, Short, C & Heady, E O (1981) The economic feasibility of crop residues as auxiliary fuel in coal-fired power plants. *American Journal of Agricultural Economics,* **63** (4), 636–644.

Essien, D & Pyle, D L (1982) Energy conservation in ethanol production by fermentation. In: Vogt, F (Ed) *Energy conservation and use of renewable energies in the bio-industries 2,* 273. Oxford: Pergamon Press.

Fairbairn, C B, Smith, A E & Hart, R (1971) Brassica waste as a feed for beef animals and dairy cows. *ADAS Quarterly Review, 3, 134–140.*

Fairgrieve, R (1979) *A policy for forestry.* London: Conservative Political Centre.

Fallows, S J & Wheelock, J V (1982) An analysis of the efficiency of the UK food system. *Agricultural Systems,* **9,** 227–236.

Fluck, R C (1981) Net energy sequestered in agricultural labour. *Transactions of the American Society of Agricultural Engineers,* **24,** 1449–1455.

Forestry Commission (1978) *The wood production outlook in Britain – a review.* Edinburgh: Forestry Commission.

Franz, R, Erckel, R, Riehm, T H, Woernle, R & Deger, H M (1983) Lignocellulose saccharification by HF. In: Strub, A, Chartier, P & Schleser, G (Eds) *Energy from biomass,* 873–878. London: Applied Science Publishers.

General Technology Systems Ltd (1981) *FAST project – biomass and regions task 1 energy from biomass technologies: state of the art review.* Report to the Commission for the European Communities. Brentford: General Technology Systems Ltd.

Hall, D O (1982) Biomass for energy in Europe. *Biomass,* **2** (4), 239–244.

Hall, D O (1983) Energy from biomass in Europe: the 1982 situation. In: Strub, A, Chartier, P & Schleser, G (Eds) *Energy from biomass,* 51–62. London: Applied Science Publishers.

Harris, G S, Leamy, M L, Fraser, T, Dent, J B, Brown, W A N, Earl, W B, Fookes, T W & Gilbert, J (1979) *The potential of energy farming for transport fuels in New Zealand.* New Zealand Energy Reserach and Development Committee, report No 46. Auckland: New Zealand Energy Research and Development Committee.

Hawkes, F R, Floyd, J R S & Hawkes, D L (1981) Operation of a laboratory-scale plug flow type of digester on pig manure. In: Palz, W, Chartier, P & Hall, D O (Eds) *Energy from biomass,* 398–405. London: Applied Science Publishers.

Heijnen, J (1983) Development of a high rate fluidised bed biogas reactor. In: Strub, A, Chartier, P & Schleser, G (Eds) *Energy from biomass,* 581–590. London: Applied Science Publishers.

Held, W & König, A (1983) Biogas driven engine. In: Strub, A, Chartier, P & Schleser, G (Eds) *Energy from biomass,* 511–516. London: Applied Science Publishers.

Hertzmark, D, Flaim, S, Ray, D & Parvin, G (1980) Economic feasibility of agricultural alcohol production within a biomass system. *American Journal of Agricultural Economics,* **62,** 965–971.

Hitzhusen, F J & Abdallah, M (1980) Economics of electrical energy generation from crop residue combustion with high sulfur coal. *American Journal of Agricultural Economics,* **62,** 416–425.

House of Commons Agriculture Committee (1980). *Economic, social and health implications for the United Kingdom of the Common Agricultural Policy on milk and dairy products.* London: HMSO.

Jenkins, B M, Horsfield, B C, Dobie, J B & Miller, G (1981) Agricultural residues: renewable energy for utility power companies. *Transactions of the American Society of Agricultural Engineers,* **24**, 197–201.

Kaufmann, R, Sutter, K & Wellinger, A (1982) Low temperature methane generation from animal manure. In: Vogt, F (Ed) *Energy conservation and use of renewable energies in the bio-industries 2,* 404–417. Oxford: Pergamon Press.

Keville, B J (1982) Short rotation forestry harvester chipper. In: Grassi, G & Palz, W (Eds) *Energy from biomass,* 83–89. Dordrecht, Holland: D Reidel Publishing Company.

Kleinhanss, W & Meister, H (1983) Economic aspects of biogas production from feed plants. In: Strub, A, Chartier, P & Schleser, G (Eds) *Energy from biomass,* 626–634. London: Applied Science Publishers.

Kofoed Nielsen, H & Nielsen, H (1983) Experiments with gas scrubbers as heat exchangers in straw furnaces. In: Strub, A, Chartier, P & Schleser, G (Eds) *Energy from biomass,* 479–483. London: Applied Science Publishers.

Kopiske, G & Eggersgluß, H (1983) Experiences with a low temperature biogas plant. In: Strub, A, Chartier, P & Schleser, G (Eds) *Energy from biomass,* 591–594. London: Applied Science Publishers.

Large, R V (1973) Factors affecting the efficiency of protein production by populations of animals. In: Jones, J G W (Ed) *The biological efficiency of protein production.* Cambridge: Cambridge University Press.

Larkin, S B C, Morris, R M, Noble, D H & Radley, R W (1981) *Resource mapping of agricultural wastes and residues.* Report on a study conducted on behalf of the Department of Energy, Energy Technology Support Unit. Silsoe: National College of Agricultural Engineering.

Lawson, G J & Callaghan, T V (in press) Primary productivity and the prospects for biofuels in the UK. *International Journal of Biometeorology.*

Lawson, G J, Callaghan, T V & Scott, R (1980) *Natural vegetation as a renewable energy resource in the UK.* Report to the Department of Energy, Energy Technology Support Unit. Grange-over-Sands: Institute of Terrestrial Ecology.

Lawson, G J, Callaghan, T V & Scott, R (1983) Biofuels from natural vegetation in the UK – the management of fuel energy crops. In: Strub, A, Chartier, P & Schleser, G (Eds) *Energy from biomass,* 212–221. London: Applied Science Publishers.

Leach, G (1976) *Energy and food production.* Guildford: IPC Ltd.

Lettinga, G (1981) Starting up of a thermophilic anaerobic digestion. In: Chartier, P & Palz, W (Eds) *Energy from biomass,* 126–130. Dordrecht, Holland: D Reidel Publishing Company.

Lewis, C W (1980) Energy considerations of biofuels production. In: San Pietro, A (Ed) *Biochemical and photosynthetic aspects of energy production,* 209–226. London: Academic Press Inc. (London) Ltd.

Linneborn, J & Tremus, J (1981) Mobile pyrolysis plant for the production of charcoal and condensable hydrocarbons from biomass. In: Chartier, P & Palz, W (Eds) *Energy from biomass,* 176–180. Dordrecht, Holland: D Reidel Publishing Company.

Lloyd, A G (1980) Energy and Australian agriculture. *Journal of the Australian Institute of Agricultural Science,* **46,** 164–169.

Lockeretz, W (1982) On-farm fuel alcohol production: economic considerations and implications for farm management. *Energy in Agriculture,* **1,** 171–184.

Loewer, O J, Black, R J, Brook, R C, Ross, I J & Payne, F (1982) Economic potential of on-farm biomass gasification for corndrying. *Transactions of the American Society of Agricultural Engineers,* **25,** 779–784.

Lucas, N G (1978) Whole crop harvesting. *Power Farming,* August–October, 1978.

MAFF (1975) *Food from our own resources:* Cmnd 6020. London: HMSO.

MAFF (1979) *Farming and the nation.* Cmnd 7458. London: HMSO.

MAFF (1980) *Agricultural and horticultural returns – final results of the June 1979 census.* Guildford: Ministry of Agriculture, Fisheries and Food.

MAFF (1983) *Annual review of agriculture 1983.* Cmnd 8804. London: HMSO.

Massey Ferguson (1981) *Massey Ferguson 200.* Brochure on MF 200 trailers. Coventry: Massey Ferguson (UK) Ltd.

Massey Ferguson (1982) *Massey Ferguson 124.* Brochure on MF 124 high output baler. Coventry: Massey Ferguson (UK) Ltd.

McLain, H D (1983a) The development of a harvester (patent pending) for 2-3 year old willow coppice. In: Strub, A, Chartier, P & Schleser, G (Eds) *Energy from biomass,* 225–229. London: Applied Science Publishers.

McLain, H D (1983b) A study of a proposed planting, growing and harvesting regime for willow biomass. In: Strub, A, Chartier, P & Schleser, G (Eds). *Energy from biomass,* 241–246. London: Applied Science Publishers.

Meinhold, K & Kögl, H (1983) Erzeugung und Verwendung von Biomasse fur Energie und Nahrung in Europa. In: Strub, A, Chartier, P & Schleser, G (Eds) *Energy from biomass,* 23–35. London: Applied Science Publishers.

Ministere de l'agriculture, Paris (1969) Perspectives à long terme de l'agriculture Française. Summary in: *Current agricultural proposals for Europe.* London: Federal Trust for Education and Research.

Mitchell, C P (1982) An experimental study of short rotation forestry for energy. In: Grassi, G & Palz, W (Eds) *Energy from biomass,* 56–61. Dordrecht, Holland: D Reidel Publishing Company.

Mitchell, C P & Matthews, J D (1981) Forest biomass as a source of energy in the UK – the potential and the practice. In: Palz, W, Chartier, P & Hall, D O (Eds) *Energy from biomass,* 239–243. London: Applied Science Publishers.

Mitchell, C P & Matthews, J D (1983) A study of short rotation forestry for energy in Britain. In: Strub, A, Chartier, P & Schleser, G (Eds) *Energy from biomass,* 181–185. London: Applied Science Publishers.

Mitchell, C P, Brandon, O H, Bunce, R G H, Barr, C J, Whittaker, H A, Tranter, R B, Downing, P, Pearce, M L (1983) Land availability for production of wood for energy in Great Britain. In: Strub, A, Chartier, P & Schleser, G (Eds) *Energy from biomass,* 159–163. London: Applied Science Publishers.

Morgan, K E (1969) Patent 4081 of 1816. *The Esso Farmer,* **21** (2), 18–22.

Mutch, W E S & Hutchinson, A R (1980) *The interaction of forestry and farming.* East of Scotland College of Agricultural Economics and Management series, No 2. Edinburgh: Department of Forestry and Natural Resources, University of Edinburgh.

National Farmers Union of Scotland (1979) *Agriculture and forestry: a policy statement.* Edinburgh: National Farmers Union of Scotland.

Neenan, M, Lyons, G & McCarthy, R (1982) The production of energy from short rotation forestry. In: Grassi, G & Palz, W (Eds) *Energy from biomass,* 77–82. Dordrecht, Holland: D Reidel Publishing Company.

Newman, D R (1979) *Aquatic plants as an energy crop in the UK.* Report to the Department of Energy, Energy Technology, Support Unit. Sheffield: University of Sheffield.

Nix, J (1982) *Farm management pocketbook* – twelfth edition. Ashford: Wye College School of Rural Economics.

Ofoli, R Y & Stout, B A (1981) Ethyl alcohol production for fuel: energy balance. In: Palz, W, Chartier, P & Hall, D O (Eds) *Energy from biomass,* 354–359. London: Applied Science Publishers.

Parisi, F (1983) Biological conversion of ethanol and alcohol routes. In: Strub, A, Chartier, P & Schleser, G (Eds) *Energy from biomass,* 803–816. London: Applied Science Publishers.

Pearce, D (1980) Energy conservation and official UK energy forecasts. *Energy Policy,* **8,** 245–248.

Pearce, D (1981) World energy demand and crude oil prices to the year 2000. *Journal of Agricultural Economics,* **32,** 341–354.

Pearce, M L (1983) Coppiced trees as energy crops. In: Strub, A, Chartier, P & Schleser, G (Eds) *Energy from biomass,* 137–141. London: Applied Science Publishers.

Pellecuer, B (1983) Valorisation de la biomasse sèche sur les exploitations agricoles – exemple d'un réseau-pilote de référence et de demonstration en région Centre. In: Strub, A, Chartier, P & Schleser, G (Eds) *Energy from biomass,* 359–363. London: Applied Science Publishers.

Picken, D J (1983) Biogas in engines. In: Strub, A, Chartier, P & Schleser, G (Eds) *Energy from biomass,* 372–375. London: Applied Science Publishers.

Pile, R S, Badger, P C & Nelson, R G (1981) *A small-scale ethanol production facility design and performance.* Paper number 81–3567 presented at a meeting of the American Society of Agricultural Engineers, Chicago, December, 15–18.

Pimentel, D, Hurd, L E, Bellotti, A C, Forster, M J, Oka, I N, Sholes, O D & Whitman, R J (1973) Food production and the energy crisis. *Science,* **182,** 443–449.

Plaskett, L G (1980) *The generation of gases from biomass by anaerobic digestion and their practical utilisation as fuel.* Report to the Department of Energy, Energy Technology Support Unit. Tiverton: Biotechnical Processes Ltd.

Radley, R W & Garrod, Y B (1981) *Vegetable oils as fuel for diesel engines.* Paper presented at an ITDG Ltd/ITIS Seminar, London, April.

Reed, T B (1981) The combustion pyrolysis, gasification and liquefaction of biomass. In: Palz, W, Chartier, P & Hall, D O (Eds) *Energy from biomass,* 496–508. London: Applied Science Publishers.

Renborg, U (1980) Energy analysis of agriculture: biology or economics – a survey of approaches, problems and traps. In: Johnson, G & Maunder, A (Eds) *Rural change – the challenge for agricultural economists,* 321–341. Farnborough: Gower.

Revuz, B & Daujat, F (1983) Production d'éthanol à partir de biomasse: technologies et bilans énergétiques. In: Strub, A, Chartier, P & Schleser, G (Eds) *Energy from biomass,* 817–821. London: Applied Science Publishers.

Rijkens, B A (1981) A novel process for the anaerobic digestion of solid wastes leading to biogas and a compost like material. In: Palz, W, Chartier, P & Hall, D O (Eds) *Energy from biomass,* 435–439. London: Applied Science Publishers.

Ritson, C (1980) *Self-sufficiency and food security.* CAS paper 8. Reading: Centre for Agricultural Strategy.

Roberts, I, Hall, N & Lipton, H (1980) The economics of fuel ethanol from crops. *Quarterly Review of the Rural Economy,* **2,** 181–185.

Rose, D W (1977) Cost of producing energy from wood in intensive cultures. *Journal of Environmental Management,* **5,** 23–35.

Sanchez, P A (1976) *Properties and management of soils in the tropics.* New York: John Wiley & Sons, Inc.

Scantland, D A, McClure, T A & Lipinsky, E S (1979) *Carbohydrate crops as a renewable resource for fuels production, Volume 2: identification of key policy issues, alternatives, and implications relating energy from biomass.* Report to the US Department of Energy. Columbus: Battelle.

Scheller, W A (1983a) Commercial experience with a plug flow anaerobic digester for the production of biogas from agricultural and food processing wastes. In: Strub, A, Chartier, P & Schleser, G (Eds) *Energy from biomass,* 492–496. London: Applied Science Publishers.

Scheller, W A (1983b) Current advances in implementing gasohol as a US automotive fuel. In: Strub, A, Chartier, P & Schleser, G (Eds) *Energy from biomass,* 964–968. London: Applied Science Publishers.

Schuchardt, F (1983) Heat production by composting animal wastes. In: Strub, A, Chartier, P & Schleser, G (Eds) *Energy from biomass,* 470–474. London: Applied Science Publishers.

Sheppard, W J & Lipinsky, E S (1983) Chemicals from biomass. In: Strub, A, Chartier, P & Schleser, G (Eds) *Energy from biomass,* 63–71. London: Applied Science Publishers.

Silvey, V (1978) The contribution of new varieties to increasing cereal yields in England and Wales. *Journal of the National Institute of Agricultural Botany,* **14,** 367–386.

Slesser, M, Lewis, C & Edwardson, W (1977) Energy systems analysis for food policy. *Food Policy,* **2,** 123–129.

Smith, A (1980) *A review of bioconversion routes cellulosic biomass to liquid fuels.* Report to the Department of Energy, Energy Technology Support Unit. Reading: University of Reading.

Smith, D L O, Rutherford, I & Radley, R W (1976) *Straw production, collection and utilisation: some energy considerations.* Paper presented at the annual conference of the Institution of Agricultural Engineers, London, May 13.

Smith, M G H (1980) Farm domestic heating. In: Staniforth, A R (Ed) *Report on 'straw as a fuel' conference.* Oxford: Ministry of Agriculture, Fisheries and Food.

Spedding, C R W & Carruthers, S P (1980) *Experimental studies on catch fuel crops.* Report to the Department of Energy, Energy Technology Support Unit. Reading: University of Reading.

Spedding, C R W, Walsingham, J M, McDougall, V D, Shiels, L A & Carruthers, S P (1979a) *The potential of photosynthetically produced organic matter as an energy feedstock.* Report to the Commission of the European Communities. Reading: University of Reading.

Spedding, C R W, Bather, D M & Shiels, L A (1979b) *An assessment of the potential of UK agriculture for producing plant material for use as an energy source.* Report to the Department of Energy, Energy Technology Support Unit. Reading: University of Reading.

Spedding, C R W, Bather, D M & Shiels, L A (1980) *An assessment of the potential of UK agriculture for producing plant material for use as an energy source.* Report to the Department of Energy, Energy Technology Support Unit. Reading: University of Reading.

Spedding, C R W, Walsingham, J M & Hoxey, A (1981) *Biological efficiency in agriculture.* London: Academic Press.

Stafford, D A (1982) Energy from farm wastes and biomass. In: Vogt, F (Ed) *Energy conservation and use of renewable energies in the bio-industries 2,* 626–633. Oxford, UK: Pergamon Press.

Stafford, D A & Etheridge, S P (1982) Farm wastes, energy production and the economics of farm anaerobic digesters. In: Hughes, D E, Stafford, D A, Wheatley, B I, Baader, W, Lettinga, G, Nyns, E J, Verstraete, W & Wentworth, B L (Eds) *Anaerobic digestion 1981,* 255–268. Amsterdam: Elsevier Publishing Company.

Staniforth, A R (1982) *Straw for fuel, feed and fertiliser?* Ipswich: Farming Press.

Stewart, D J (1980) Energy crops to methane. In: Stafford, D A, Wheatley, B I & Hughes, D E (Eds) *Anaerobic digestion.* London: Applied Science Publishers.

Stewart, G A, Rawlins, W H M, Quick, G R, Begg, J E & Peacock, W J (1981) Oilseeds as a renewable source of diesel fuel. *Search,* **12** (5), 107–115.

Stott, K G, Parfitt, R I, McElroy, G & Abernethy, W (1983) Productivity of coppice willow in biomass trials in the UK. In: Strub, A, Chartier, P & Schleser, G (Eds) *Energy from biomass,* 230–235. London: Applied Science Publishers.

Teisen, P H (1982) The use of wood and straw-fired boilers in agricultural production. In: Vogt, F (Ed) *Energy conservation and use of renewable energies in the bio-industries 2,* 640–648. Oxford: Pergamon Press.

Thring, M W & Crookes, R J (1981) Farm straw as fuel for power. In: Palz, W, Chartier, P & Hall, D O (Eds) *Energy from biomass,* 516–522. London: Applied Science Publishers.

Tjernshaugen, O (1981) The manure cellar as a source of heat. In: Palz, W, Chartier, P & Hall, D O (Eds) *Energy from biomass,* 150–155. London: Applied Science Publishers.

Tougaard Pedersen, T (1982) Heat from waste products especially straw and manure by aerobic combustion. In: Robinson, D W & Mollan, R C (Eds) *Energy management and agriculture.* Dublin: Royal Dublin Society.

Tranter, R B (1983) Land. In: Spedding, C R W (Ed) *Fream's agriculture.* Chapter 6. London: John Murray.

Trenbath, B (1974) Biomass producitivity of mixtures. *Advances in Agronomy,* **26,** 177–210.

Van Swaaij, W P M (1981) Gasification – the process and the technology. In: Palz, W, Chartier, P & Hall, D O (Eds) *Energy from biomass,* 485–495. London: Applied Science Publishers.

Van Swaaij, W P M & Beenackers, A A C M (1982) The methanol from biomass pilot plant programme of the Commissions of the European Communities. In: Grassi, G & Palz, W (Eds) *Energy from biomass,* 213–218. Dordrecht, Holland: D Reidel Publishing Company.

Wareing, P F & Allen, E J (1977) Physiological aspects of crop choice. *Philosophical Transactions of the Royal Society,* London B, **281,** 107–119.

Webb, M & Pearce, D (1975) The economics of energy analysis. *Energy Policy,* **3,** 318–331.

Webb, S B (1981) The impact of increased alcohol production on agriculture: a simulation study. *American Journal of Agricultural Economics,* **63** (3), 532–537.

White, D J (1980) The potential uptake of straw as fuel in agriculture and horticulture. In: Staniforth, A R (Ed) *Report on 'straw as a fuel' conference.* Oxford: Ministry of Agriculture, Fisheries and Food.

Wilkins, R J, Newton, J E, James, P J & Walsingham, J M (1981) Possibilities for change in the output from grassland: an overall view. In: Jollans, J L (Ed) *Grassland in the British economy.* CAS Paper 10. Reading: Centre for Agricultural Strategy.

Willey, R W (1979) Intercropping – its importance and research needs: part 1. Competition and yield advantages. *Field Crop Abstracts,* **32** (1), 1–10.

Wilson, H T (1982) Status report on the FWPP biomass gasification test facility. In: Grassi, G & Palz, W (Eds) *Energy from biomass,* 196–200. Dordrecht, Holland: D Reidel Publishing Company.

Wonder, B & Simpson, D (1982) Economics of large-scale and on-farm production of fuel from crops in Australia. *Energy in Agriculture,* **1,** 155–170.

Wright, D J (1975) The natural resource requirements of commodities. *Applied Economics,* **7,** 31–39.

Zerbin, W O (1983) Generating electricity with Imbert gasifier. In: Strub, A, Chartier, P & Schleser, G (Eds) *Energy from biomass,* 436–442. London: Applied Science Publishers.

Index

Centre Publications

Reports

1 *Land for agriculture* (1976) £1.50
2 *Phosphorus: a resource for UK agriculture* (1978) £1.75
3 *Capital for agriculture* (1978) £2.75
4 *Strategy for the UK dairy industry* (1978) £2.95
5 *National food policy in the UK* (1979) £2.85
6 *Strategy for the UK forest industry* (1980) £8.50
7 *The efficiency of British agriculture* (1980) £2.85

Papers

1 Marsh, J S (1977) *UK agricultural policy within the European Community* £1.50
2 Tranter, R B (Ed) (1978) *The future of upland Britain* £13.50
3 Harrison, A, Tranter, R B & Gibbs, R S (1977) *Landownership by public and semi-public institutions in the UK* £1.75
4 Collins, E J T (1978) *The economy of upland Britain 1750–1950: an illustrated review* £2.20
5 McCalla, A F (1978) *International agricultural research: potential impact on world food markets and on UK agricultural strategy* £1.50
6 Swinbank, A (1978) *The British interest and the green pound* £1.50
7 Robbins, C J (Ed) (1978) *Food, health and farming: a report of panels on the implications for UK agriculture* £2.40
8 Ritson, C (1980) *Self-sufficiency and food security* £2.00
9 Tranter, R B (Ed) (1981) *Smallfarming and the Nation* £2.00
10 Jollans, J L (Ed) (1981) *Grassland in the British economy* £20.00
11 Marshall, B J & Tranter, R B (Eds) (1982) *Smallfarming and the rural Community* £2.50
12 Hallam, D (1983) *Livestock development planning: a quantitative framework* £10.00
13 Carruthers, S P & Jones, M R (1983) *Biofuel production strategies for UK agriculture* £6.50

Mailing list: To receive notification of future publications please ask to be put on the mailing list.

Standing orders: To receive invoiced copies of future publications please ask to be put on the standing order list.

Orders: All publications available from: The Centre for Agricultural Strategy, University of Reading, 1 Earley Gate, Reading RG6 2AT. Prices quoted include postage.